ROOTED IN TIME

CAROLE T. GEE

Illustrated by

CHANNING REDFORD

ROOTED IN TIME

Living Fossils and Other Tenacious Plants

Johns Hopkins University Press

Baltimore

Printed in the United States of America on acid-free paper
9 8 7 6 5 4 3 2 1

Johns Hopkins University Press
2715 North Charles Street
Baltimore, Maryland 21218
www.press.jhu.edu

Library of Congress Cataloging-in-Publication Data
Names: Gee, Carole T., author. | Redford, Channing, illustrator.
Title: Rooted in time : living fossils and other tenacious plants/ Carole T. Gee ; illustrated by Channing Redford.
Description: Baltimore : Johns Hopkins University Press, [2024] | Includes bibliographical references.
Identifiers: LCCN 2023049867 | ISBN 9781421449388 (hardcover) | ISBN 9781421449395 (ebook)
Subjects: LCSH: Paleobotany.
Classification: LCC QE905 .G44 2024 | DDC 561—dc23/eng/20240223
LC record available at https://lccn.loc.gov/2023049867

A catalog record for this book is available from the British Library.

For my late father, Peter M. Gee,
and my mother, Ann Yin Gee,
who cheerfully raised me as a nature lover,
thereby setting me on my life's path
as a professional botanist, research
paleobotanist, and soft-rock geologist—
with my most loving thanks

—CAROLE

For Jim Braddock

—CHANNING

CONTENTS

Foreword, by Sir Peter Crane, FRS ix

Preface: The Making of a Plant Lover xvii

PART I

1 Living Fossils: Morphological Look-Alikes, Tenacious Survivors, and Relict Members of Ancient Lineages 3

2 Cyanobacteria and Stromatolites: The Toughest and Longest-Lived Green Survivors 17

PART II

3 Standing Tall 33

4 Clubmosses & Co. 45

5 The Horsetail, or Scouring Rush, *Equisetum* 61

6 Ferns and Tree Ferns 75

PART III

7 Bearing Seeds and Woody Cones 89

8 Cycads 101

9 The Maidenhair Tree, *Ginkgo* 115

10 Araucarias, Kauris, and the Wollemi Pine 131

11 Podocarps 149

12 The Dawn Redwood, *Metasequoia* 161

13 The Japanese Umbrella Pine, *Sciadopitys* 175

PART IV

14 Coming into Flower 191

15 The Water Lilies *Nymphaea* and *Nuphar* 201

16 The Sacred Lotus, *Nelumbo* 217

17 The Mangrove Palm, *Nypa* 231

Index 245

FOREWORD

Time is the great unifier: everything embodies history. But too often our view of the past embraces only human history: the comings and goings of people and empires. More profound, and more consequential for understanding ourselves and other living species, is the deep history of our planet: the comings and goings of continents and the plants and animals that were here long before *Homo sapiens*. In this book Carole Gee expands our perspective by going broad and deep, inviting us to explore the variety of plant life and travel back into the billion-year history of the world of plants, and, through that, the greening of our planet.

In an increasingly urbanized world, seemingly remote from nature, it is too easy to overlook our fundamental reliance on plants, but none of us can opt out of our dependence on the alchemy of photosynthesis that transforms energy from the sun into the chemical energy that powers the ecosystems of which we are part. Water and carbon dioxide are the simplest of materials, but through its biomolecular magic photosynthesis produces the complex energy-rich sugars that sustain life. Photosynthesis also produces oxygen, a potent by-product, which in the deep past transformed our planetary atmosphere and created the conditions in which we and the life we know can flourish.

The rise of plants changed our planet in other ways. In the oceans, more complex and more abundant plants were a new source of energy for new kinds of animals to exploit. They provided the foundation for new kinds of marine ecosystems, a process that continued as plants then began to colonize freshwater lakes, ponds, and rivers. From freshwater, plants moved onto the land and in the process helped stabilize the earth's surface by reducing erosion and contributing to the build-up of soil. And soon, with a new source of unexploited food on land, it is no surprise that land animals, especially arthropods, followed.

As plants grew larger, and more complex, they also became entrained in ever more subtle and ever more complex interactions with animals. Plants became not just a source of nutrition, but also places of refuge and sources of new kinds of chemicals that inserted themselves in new ways into animal biology. Through the process of evolution plants have become creative biochemists. Plants are also the ultimate ecosystem engineers. They created new habitats that could nurture and support new kinds of plants and animals.

Over time, the worlds of plants and the worlds of animals became ever more closely intertwined, and human society is no exception. By virtue of their abundance and variety, plants have become essential in all of our lives. Hundreds of plant species contribute to the human diet, from our most important crop plants to the myriad of minor herbs and spices that enliven our palate. In addition, plants are fundamental to the human pharmacopeia, and they provide us with timber, fiber, and many other raw materials. Plants are also unmistakably embedded in human culture: witness their prominent place in diverse religions and their inspiration for art of all kinds.

The chapters in this book take us on a tour of some of our author's favorite plants, introducing them in all their evolutionary and cultural fascination. A particular focus is "living fossils," plants with special connections to the deep history of plant life. These exemplars provide key points of reference that help us navigate the bewildering variety of plant life and explore plant evolution. Through these examples we are also introduced to the fossil evidence for plants of the past and what fossils tell us about the origins of the modern world.

A common misconception is that plants have a poor fossil record. After all, plants have no bones, no teeth or other hard parts, that might predispose them to be preserved in rocks laid down by rivers or in lakes in the distant past. In fact, while the plant fossil record has limitations, it still has much to tell us. Its limitations are merely different from those of the animal fossil record. The most problematic feature of the plant fossil record is that complex plants are rarely preserved as whole organisms. Leaves, spores, seeds, and other plant parts are generally shed separately. Much of the work of plant paleontologists is trying to figure out which plant parts might have come from the same living species. But looked for in the right places, and studied intensively using the right techniques, the fossil record of plants is both rich and informative, as we learn in the pages of this book.

Spores and pollen grains, while tiny, have surprisingly resistant walls and can be extracted in vast quantities from ancient sediments deposited on the land surface, as well as in marine rocks deposited close to shore. Leaves, produced in their thousands on a single living tree and regularly shed during the cycles of growth, are also common in the fossil record and give clues to ancient climates. Fossil fruits and seeds, and even flowers, the most delicate and ephemeral of all plant parts, can be common and beautifully preserved in certain kinds of rocks if searched for in the right way. This book points to the wonder of fossil plants. It provides a way into the world of paleobotany.

The story of plant evolution begins very early in the history of the earth, as recounted in the first section of this book. Discoveries from the first four billion years of our planet's history provide evidence of the plantlike bacteria that flourished in the first oceans. Their close relatives became incorporated into the cells of the true plants that came later. By around a billion years ago fossils, as well as age estimates based on the DNA sequences of living organisms, suggest that green algae, the first true plants, had already evolved. Somewhat later they are also known from rocks deposited not in the oceans, but in freshwater showing that some of these early green plants were beginning to colonize environments closely associated with land.

The emergence of plants capable of living on land began about 450 million years ago, perhaps as freshwater green algae were forced to contend with life in air as the water in which they were living dried out. By about 400 million years ago their descendants were larger, more complex, and had acquired many of the key features essential for life on land: spores with resistant walls that could survive desiccation and waxy coverings of the plant surfaces that helped reduce water loss from internal tissues. Also crucial was the evolution of stomata, exquisite cellular valves that help regulate the exchange of water vapor, carbon dioxide and oxygen between the plant tissues and the atmosphere.

Exactly how some of these early plants on land relate to plants alive today is uncertain. Some show features of true mosses and hornworts, and as we see in the second section of this book some were clearly related to living clubmosses. Others were probably the ancestors of horsetails, while others have been implicated in the origin of early ferns. Also clear, when information from fossil and living plants is combined, is that all the plants living on land today trace their ancestry to the same group of early colonizers known as fossils from the Silurian and Early Devonian. The key transition from living in water to living on land was made by just a single lineage of plants that proved highly versatile. Among animals, colonization of the land was quite different. It was accomplished independently by arthropods, vertebrates, and many other groups at different times.

The earliest plants on land were all small, and available light was plentiful. But as plants diversified and grew taller, they began to shade their neighbors. A struggle for light was soon underway, and the origin of trees in the Middle and Late Devonian appears to have been one response. Large treelike forms evolved independently in several different plant lineages. Clubmosses, horsetails, and several extinct plant groups were able to create a substantial trunk by different innovations in their internal anatomy. One innovation was a woody trunk very similar to the trunks of most modern trees, and it was from plants with these kinds of stems that diverse groups of seed plants emerged during the Late Paleozoic and Mesozoic.

Seed plants, excluding the flowering plants or angiosperms, are the focus of third section of this book. The earliest fossil seeds are recognized at the end of the Devonian about 370 million years ago, and the plants that produced them quickly flourished and diversified. Many groups of extinct seed plants are known from the Paleozoic and Mesozoic, but today only five remain: cycads, ginkgo, conifers, the relatively obscure gnetophytes, and angiosperms. All but the angiosperms were probably in existence before the end of the Paleozoic and evidently survived the Permian-Triassic mass extinction.

One consequence of the origin of land plants from freshwater green algae is that the reproductive systems of clubmosses, ferns and horsetails require at least some free water to be effective. The male gametes, powered by flagellae, swim through the soil moisture on their way to fertilize an egg. Such reliance on water for reproduction would have constrained the habitats in which early land plants could flourish, as it does for most clubmosses, horsetails, and ferns today. The evolution of seeds, in which fertilization occurs inside a specialized structure retained on the mother plant, appears to have been one response to a strong pressure to minimize the need for free water for reproductive success.

Reproduction in seed plants is distinguished from reproduction in ferns and similar plants by several structural and developmental specializations but especially important was the evolution of a new process, pollination. For reproduction in seed plants to be successful the male gametes inside the male spores (pollen grains) have to be transferred into proximity to the egg cells inside the developing ovule. If fertilization of an egg by a male gamete is successful, the resulting zygote will develop into an embryonic plant inside the seed that develops from the ovule.

This two-stage reproductive process in which fertilization takes place inside the developing ovule and is preceded by pollination occurs in all seed plants. In addition, in conifers, angiosperms and gnetophytes, the male gametes have lost the ancestral ability to swim and instead are delivered to the egg by a pollen tube. Interestingly, ginkgo and cycads are different and exhibit an astonishing holdover from the aquatic ancestry of all seed plants. After pollination, the male gametes develop inside the ovule and are eventually released into a small fluid-filled space. Here they swim just a very short distance before fertilizing an egg.

More diverse than ginkgo and cycads, and of greater ecologically importance globally, are the conifers of which some of the most fascinating: the dawn redwood, the Wollemi pine, monkey puzzles, kauris, the umbrella pine, and podocarps are featured in this book. All have a rich history that has been illuminated by the studies of their fossil record. As for many other groups, including ginkgo, cycads and gnetophytes, a common theme is that the fossil record shows that they were once distributed much more widely. Their more limited distributions today reflect reductions in their natural range that resulted from the changes in the earth's climate over the past 66 million years.

The final section of the book introduces the angiosperms, the flowering plants, the hyper-diverse success story of the plant kingdom. Angiosperms first appear in the fossil record in the Early Cretaceous, around 130 million years ago, but quickly dominated the vegetation of many habitats in many parts of the world. In the process many ancient ferns and seed plants of the earlier Mesozoic became less important in global vegetation. Some were lost completely. But at the same time, many groups of plants, from conifers and cycads to clubmosses and ferns, evolved in new directions resulting in the forms that we are familiar with today. The rise of angiosperms resulted ultimately in an overwhelming exuberance of more than 400,000 living species with profound impacts on the diversity of insects and other animals living on land. The origin and diversification of flowering plants ushered in fundamental changes in global ecosystems and created the foundation on which the natural world of today is built.

From the flowering plants of the Cretaceous, to the giant club-mosses of the Paleozoic, and the plantlike blue-green bacteria of the Precambrian, plants have been central participants in the ever changing history of our planet. Though rarely a concern in affairs of state, the direct benefits to people provided by plants, and their crucial roles in regulating our environment, mean that plants are still central in all of our lives. Whether we recognize it or not, we ignore these simple facts at our peril. In helping us understand the variety of plant life and how it has changed through earth history this book gives us a broader perspective. The vast spans of time covered and range of plants to which we are introduced reveal a magnificent story that has unfolded over billions of years, and of which *Homo sapiens* is part. As Darwin famously said, "there is grandeur in this view of life." *Rooted in Time* invites us to appreciate and dip into that story.

Sir Peter Crane, FRS
President, Oak Spring Garden Foundation

PREFACE

The Making of a Plant Lover

I was an odd child. Instead of romping around with puppies, petting kittens, or reveling in the fluffiness of yellow chicks at Easter time, I was loving the leaves on my parents' kumquat tree. Each thick leaf was so stiff and had such a straight midrib that I was exceedingly fond of stroking the leaf's lamina down the middle. It was the same with the dark green glossy leaves of my mother's gardenia bush, though those leaves had a much softer texture that begged for more of a caress. I did not like the feel of the leaves of the fig tree in the backyard, which were so hairy that they were downright prickly, though I loved picking and eating the fig fruits when they turned soft and purple.

The flower scents around my parents' house were heady: the intoxicating perfume of gardenias, the bright, sweet fragrance of citrus blossoms, the spiciness of oleander flowers. The fruits of the warmth-loving plants in our southern California garden were also central to my early childhood experiences. I used to mix "professor drinks" of mud and water laced with the flaming red berries of firethorn or orange kumquats, which bobbed up and down merrily in the watery mix.

Sometimes I'd add some dusty powder from the old red bricks that I ground up with rocks to thicken my laboratory solutions, stirring them with twigs from the yard. Despite having five siblings, I often played by myself. I loved the solitary and contemplative work of experimentation in my outdoor "lab" to see how different materials reacted in my muddy mixtures.

My mother was the one who chose the plants at the garden shops. She was also the one who taught us their exotic names: *yucca, oleander, gardenia, bird of paradise, jade tree*. It was both my parents who planted the numerous fruit trees in their backyard, which must have reminded them of their childhood in rural southern China, where the fruits were so bountiful. In addition to kumquats, oranges, Meyer lemons, and persimmons, we had trees with apples, Chinese pears, and of course figs. Despite its hairy leaves, I loved the fig tree best, in my mind proclaiming it my tree because it produced the juicy fruits that I loved. At the peak of fig maturity, I had to outcompete the birds and ants, the avian frugivores swooping in from the air and the lines of insects marching up the trunk and swarming all over the burst-open, sun-ripened fruits. I can't say that I always won. In retrospect, I guess that this was my first experience with plant-animal interactions.

It was my mother who yearned for palm trees in our backyard. I remember going with my family to Unimart, our Target superstore of the 1970s, to select three knee-high fan palms. It was my father's job to dig in the newly bought plants and tend to their care. I remember how he planted the baby palms, digging huge holes to hold enough water for them in the dry climate of Los Angeles. These California short fan palms were a constant presence throughout my childhood, when we played in the yard pretty much every day. Now, about a half century later, they have morphed into tall palm trees (fig. P.1), swaying gracefully against the sunny, deep turquoise sky and paying tribute to yet another beautiful Los Angeles sunset. The best thing, though, is that when I see Las Tres Palmas from afar, I know that I am nearly home.

Figure P.1. Las Tres Palmas, standing tall in my mother's backyard and perpetually appearing in my dreams. Photo by author.

My memories of my friendship with the yard shrubs would have been comparable to any other child's memories of catching crayfish or hiking (although we did that too) except that my father started taking us to botanical gardens on the weekends when I was a preteen. There, my intellectual interest in plants became engaged. In the early '70s, entrance to botanical gardens was free to the public, and there are many spectacular gardens in the Greater LA area. A family of six could wander around comfortably, enjoying the manicured lawns and brightly colored flower beds. But what I liked best was the native California habitats—the aromatic shrubby chaparral vegetation and the dusty evergreen oak woodlands. Even better, at one botanical garden there was a tram that I rode over and over. On this tram ride, a guide told you what the plants were called and explained how they grew. Each guide highlighted different plants and communities, so my knowledge of California flora grew with every tram ride. These fun facts about plants must have prompted my love for plant systematics and ecology.

The first green thing of my own was an African violet with velvety purple flowers, which I grew on the windowsill in my bedroom. It was serendipitous that my room had a north-facing window, with the perfect indirect-light exposure for this type of house plant. When I discovered that my little African violet always grew toward the light and was becoming quite lopsided, I started rotating it about a quarter turn each day when I came home from school, which made its new leaves grow out more evenly. Much later, in college, I learned that this phenomenon is called *phototropism* and is common in photosynthetic organisms. Back then, though, my high school biology teacher was surprised and disbelieved my observations that the directional growth of plants toward the light could happen so quickly, within a day and not over the course of weeks. I think that this was one of my first forays into empirical science.

Thus, from my childhood I knew that I wanted to pursue the study of plants. At that time, I also learned the word for the science of plants, *botany*. The question was now, what kind of botany did I want to study? There were so many interesting fields—marine botany, native plant ecology, plant biochemistry . . . As it turned out, some random classes in geology showed me my path.

At Pomona College, I had a remarkable teacher of geology, Donald Zenger, who was as energetic and enthusiastic as they come. At that time there were only two professors of geology at Pomona, and

Figure P.2. On a pilgrimage to Shark Bay, Western Australia, where rare stromatolites grow in abundance in a hypersaline embayment. Self-photo by author.

Dr. Zenger was the soft-rock guy, covering introductory geology, historical geology, and sedimentology. It was Dr. Zenger's class in historical geology, a fast-paced romp through the history of life and the geology of the earth, that steered my attention to paleontology. I learned about so many things that I hadn't known about before—stromatolites (fig. P.2), ammonites, plate tectonics, and, best of all, that there were such things as fossil plants. That a distinctive, tongue-shaped leaf called *Glossopteris* was major evidence that the southern continents had once formed an ancient supercontinent called Gondwana was amazing to me.

The train track switches were now set to botany, and the ultimate station was paleobotany. My next stop was graduate school at the University of Texas at Austin, where I spent one of the most joyful times of my life. Everyone that I met was a botanist and deep into learning about plants. Plants were glorious in all ways, whether as algae, fungi, or vascular plants, in the lab or out in the field. It was there that I learned about their morphology, anatomy, and botanical affinities, as well as how to use a compound microscope or operate a transmission electron microscope with equal ease. My best friend in botany graduate school called us "botany geeks," but I thought it was a normal thing to love plants. I still do.

Today's academic world forces scientists to specialize early and narrowly, yet the pivotal times in my life were my childhood years of playing freely with plants in an urban garden and the years of broad, rich training in the natural sciences in college and graduate school. As a research scientist and university professor, it is easy to fall down a narrow rabbit hole in an attempt to pursue specific scientific questions. It often gets harder to see how and when the answers connect up with the rest of the natural world. This, however, is what I think most people are interested in: an understanding of nature and their role in earth's ecosystem.

In this book, I wish to introduce the reader to my favorite plants, those that first appeared millions, if not hundreds of millions, of years ago and still green the planet today with their gentle dignity and presence. I have been collecting their stories and photos from all over the world for many decades now. While some of these plants will turn out to be old friends from the garden and others will be new acquaintances, I hope that you will grow as fond of them as I have. I think that you will not only enjoy them for their quirky charms but also admire them for their quiet persistence and longevity through evolutionary time. The lineages of some plants are so ancient that they witnessed the rise and fall of the dinosaurs, while others helped power the flowering plant revolution 100 million years ago. Oh, the stories that these plants and their ancestors could tell.

In many ways, this book has been a lifetime in the making. My greatest thanks go to my father, the late Peter M. Gee, who shared his love of the great outdoors by taking his family on outings to the sea, the mountains, or the desert every weekend. My love and gratitude also go to my mother, Ann Yin Gee, who was our hands-on guide to interacting with nature. She was great at teaching us practical things, such as filling the garden with colorful plants or catching fish with a stick, a length of twine, a bite of hotdog, and a safety pin. I am also thankful to the many who have accompanied me on my plant adventures—friends, teachers, colleagues, students—and especially my family, who have shared with me everything from collecting cones of every conifer encountered to seeking out one of the last natural stands of the Japanese umbrella pine. I am grateful to paleobotanist Dr. Robert Gastaldo for his thorough review of the manuscript and his helpful suggestions. Lastly, special thanks go to science acquisitions editor Tiffany Gasbarrini and assistant acquisitions editor Ezra Rodriguez at Johns Hopkins University Press and to New York architect-cum-global botanical artist Channing Redford for their encouragement, trust, well-appreciated criticism, and patience during the writing of this book.

Happy botanizing!

PART I

CHAPTER 1

Living Fossils

Morphological Look-Alikes, Tenacious Survivors, and Relict Members of Ancient Lineages

Everyone loves the wonder of a living fossil. The thought of a horseshoe crab or ginkgo tree surviving hundreds of millions of years to the present day, or perhaps the more worrisome idea of bringing back ferocious dinosaurs, puts a smile on our faces. In fact, *living fossils* is such a highly descriptive term that we all feel that we intuitively understand its meaning: organisms that can be found deep in the fossil record yet live on today. Horseshoe crabs and ginkgoes (fig. 1.1), as well as cycads, *Nautilus*, and coelacanths, reside in our collective knowledge as tangible links to earth's flora and fauna from the geological past.

Yet, are these popular organisms truly living fossils? What exactly is a living fossil? Is it an organism found alive in today's world that has a long paleontological record? Is it a coelacanth like *Latimeria*, which has the look of an ancient, fleshy-finned, monster fish from the depths of the ocean? Does a plant like the maidenhair tree, *Ginkgo*, which has flourished on earth for hundreds of millions of years, since the Age of Dinosaurs, count? Curiously enough, a few plants and animals, such as

Figure 1.1. Fossil and extant leaf of *Ginkgo*. *Left*, a 50-million-year-old leaf of *Ginkgo adiantoides* from Washington State. Photo by G. J. Retallack, Creative Commons Attribution-ShareAlike 4.0 International license. *Right*, a leaf of the only living species of *Ginkgo*, *G. biloba*. Photo by Daniel Miłaczewsk, Creative Commons Attribution-ShareAlike 3.0 Unported license.

the dawn redwood, *Metasequoia*, and the slit snails of the Pleurotomariidae, were known and first named as fossils, thought to be long extinct, and later discovered to be still living. Surely, these plants and animals must qualify as living fossils.

Because of the descriptive nature of the term, it is easy to understand why *living fossil* conjures up the image of a living plant or animal that can be found as a fossil. Implied in this definition is some sort of close biological relationship between the fossil and a living organism. While a palm frond may look similar to the leaf of a cycad (see fig. 8.3), their foliage only bears a superficial resemblance, which gives the mistaken impression of a close relationship. Instead, cycads that live on earth today would be considered living fossils not because of any relationship to living palms but on the basis of the existence of similar-looking cycad leaves from the Mesozoic. However, are there plants and animals alive today that look exactly the same as those that we see in the rock record? If so, does morphological similarity really imply a close biological relationship? Or is there more to being a living fossil beyond a superficial resemblance? These are all questions that paleobotanists and botanists tackle today using the modern tools found in paleontology, botany, and molecular biology.

More than two centuries ago, the term *living fossil* was coined by Carl Gustav Carus (1789–1869; fig. 1.2), an eastern German naturalist, who used it in the title of his 1818 scientific paper *Beitrag zur Lehre von den lebendigen Fossilien* (Contribution to the study of living fossils). In this article, Carus reported on amphibians that were found alive after being buried deep in the earth. He thought that these animals had fallen into a type of hibernation for hundreds or even thousands of years. Yet, when he dug up the amphibians and suddenly exposed them to light and warm conditions, they died immediately. The observations and interpretations in Carus's article imply that it was then thought possible for ancient organisms in the fossil record to come back to life, albeit for a short period of time.

Figure 1.2. Portrait of Carl Gustav Carus painted by J. C. Rößler ca. 1810, approximately eight years before the publication of Carus's observations on "living fossils." This image was reproduced in a book by Georg Beutel called *Bildnisse hervorragender Dresdner aus fünf Jahrhunderten. Mit kurzen Lebensbeschreibungen*, which was published in Dresden in 1908.

Forty-one years later, English naturalist Charles Darwin (fig. 1.3) touched on the term *living fossils* in his now-famous book *On the Origin of Species through Natural Selection*, which was published in London in 1859. The definition of a living fossil must have shifted since Carus's usage, for Darwin named the platypus and the lungfish as examples of living fossils. Darwin's explanation for the persistence of these "anomalous forms" to the present day was that they had survived in areas of limited extent and thus been exposed to less severe competition (Darwin 1859, 107).

More than 120 years later, in 1984, a more thorough definition was formulated by Tom Schopf, who put forth several ways that an organism living today could qualify as a living fossil:

1. Long endurance: surviving over geological time
2. Morphological resemblance: having a morphological and physiological similarity to another species over a long period of geological time

Figure 1.3. Photograph of Charles Darwin, believed to have been taken in 1854, five years before the publication of his seminal book on evolution. This photo is the frontispiece of his son Francis Darwin's book *The Life and Letters of Charles Darwin*, published in London in 1887.

3. Ancient traits: possessing many primitive morphological features
4. Refuge distribution: found in a relict distribution and also meeting one of the three conditions above
5. First thought extinct, later found alive: previously believed to have died out completely
6. Low diversity: having low taxonomic diversity and meeting one or more of the first three conditions above

Schopf's definitions offer us a firm basis on which to decide whether plants can be called living fossils. Follow-up studies published since then have expanded this list to include several more criteria, including a slow rate of evolutionary change, low extinction and speciation rates, presence of characteristics that appear to be ancestral, and genealogical divergence in the very distant past. The addition of these criteria gives us even a wider latitude for determining which plants could be considered to be living fossils. Of course, one of the most fundamental determinants is a plant's level of taxonomic hierarchy. Are living fossils only known on the species level, or can we include higher taxa and lineages as well? *Rooted in Time* takes a more expansive and inclusive approach, embracing plants of broad lineages as well as those of more narrowly defined genera and species. This approach encompasses major phyla such as cyanobacteria, genera such as *Nypa*, the mangrove palm, and species such as *Ginkgo biloba*.

In recent years, some scholars have been dissatisfied with what they perceive to be the validity of the concept of a *living fossil*. Arguments have been made for the retirement of this term, but just as many scientists have called for refining or redefining the concept to reframe research questions and identifying phenomena to be explained. Others want to expand the concept by using various working definitions in different research contexts, while some worry that the label *living fossil* is not positive but pejorative, similar to referring to someone who refuses to accept new things as a "dinosaur." It is clearly a tumultuous time for the term *living fossil*, as it is being actively debated whether the concept is solid and valuable or, in the face of emerging fields such as molecular biology, needs to be revamped or even abandoned. However, despite the deep biological and philosophical arguments pro and con at this time, the jury is still out. Because it is beyond the goals of this chapter to enter into the current fray, Schopf's 1984 list of useful definitions will be consulted in deliberations on which plant lineages do or do not constitute living fossils.

Central to the concept of a living fossil is the depth of paleontological time. Paleontological time is based on a geological timeline, which documents geological events on earth (fig. 1.4). Most geological time charts start with the formation of the earth 4.5 billion, or 4,567 million, years ago, at the start of the Precambrian. This major event for our planet was followed by the formation of the oceans, then the appearance of life in these waters, which is generally accepted as occurring 3.8 billion, or 3,800 million, years ago. The first cyanobacterial cells and microbial mats appear 3.5 billion years ago (chapter 2). It was the cyanobacteria, along with other microorganisms, that made up the greater part of the life during the Precambrian.

The next key events in the evolution of plants occurred in the Phanerozoic eon, which is divided into three eras—the Paleozoic, the Mesozoic, and the Cenozoic. Many major events, such as the conquering of land, the evolution of arborescent forms, the formation of forests, and the advent of wood and seeds, all occurred during the Paleozoic. A number of plants discussed in this book have roots that reach deep into the Paleozoic, such as the clubmosses (chapter 4), *Equisetum* (chapter 5), ferns and tree ferns (chapter 6), cycads (chapter 8), and *Ginkgo* (chapter 9). The oldest relatives of conifers, such as araucarias (chapter 10) and podocarps (chapter 11), first appeared a little later, in the early Mesozoic. The last major key event in plant evolution, the rise of the angiosperms, also known as flowering plants, occurred in the Early Cretaceous, the second-to-last epoch of the Mesozoic. By the end-Mesozoic, all major groups of flowering plants, including water lilies (chapter 15), lotuses (chapter 16), and palms (chapter 17), had emerged. At this time, there was an enormous shift in global vegetation from the previously gymnosperm-dominated flora to our modern angiosperm-filled flora. In the Cenozoic, the flowering plants grew enormously in abundance and diversity, dominating many habitats ecologically.

Thus, the paleontological record of the living fossils discussed in this book covers a huge span of time, from the Precambrian 3.5 billion years ago to the present. Each plant bears a close morphological resemblance to its ancient relatives. Some plants are so similar on the generic level that they are sometimes given the genus names of their living sister plants, such as *Equisetum* and *Araucaria*. Living and ancient royal ferns of the family Osmundaceae are so remarkably similar that even the micrometer-scale size of their cell nuclei is the same. The plants portrayed in this book all possess "primitive," or ancestral, characters that make them stand out in the plant kingdom.

Several plants and plant groups discussed here have relictual distributions, especially when compared with their broader distributions in the past. For example, stromatolites—calcareous mounds made up of layers of cyanobacteria, other microbes, and trapped sediment—once flourished along shorelines all over the world. They were especially abundant before the advent of metazoan grazers in the Cambrian explosion 539 million years ago. Today, stromatolites are only found in a few hypersaline bays (fig. 1.5) and some inland lakes (see chapter 2).

Among Mesozoic plants, the *Araucaria* conifer was one of the dominant trees in forests across the globe throughout the Jurassic but today are found naturally in dwindling numbers in Australasia and in South America. In the Triassic and the Jurassic, fossil *Ginkgo* leaves could be found throughout the Northern Hemisphere; however, the genomes of the today's last surviving species of *Ginkgo*, *G. biloba*, can be traced back to only three refugia in China. In the past, the Japanese umbrella pine, *Sciadopitys*, also had a much broader distribution in the Northern Hemisphere and was even a dominant swamp tree along the Proto–North Sea coastline of Germany during the Miocene. Today, however, it is endemic to a small patch of temperate rain forest near Osaka in Japan. Finally, from the Upper Cretaceous to the Eocene,

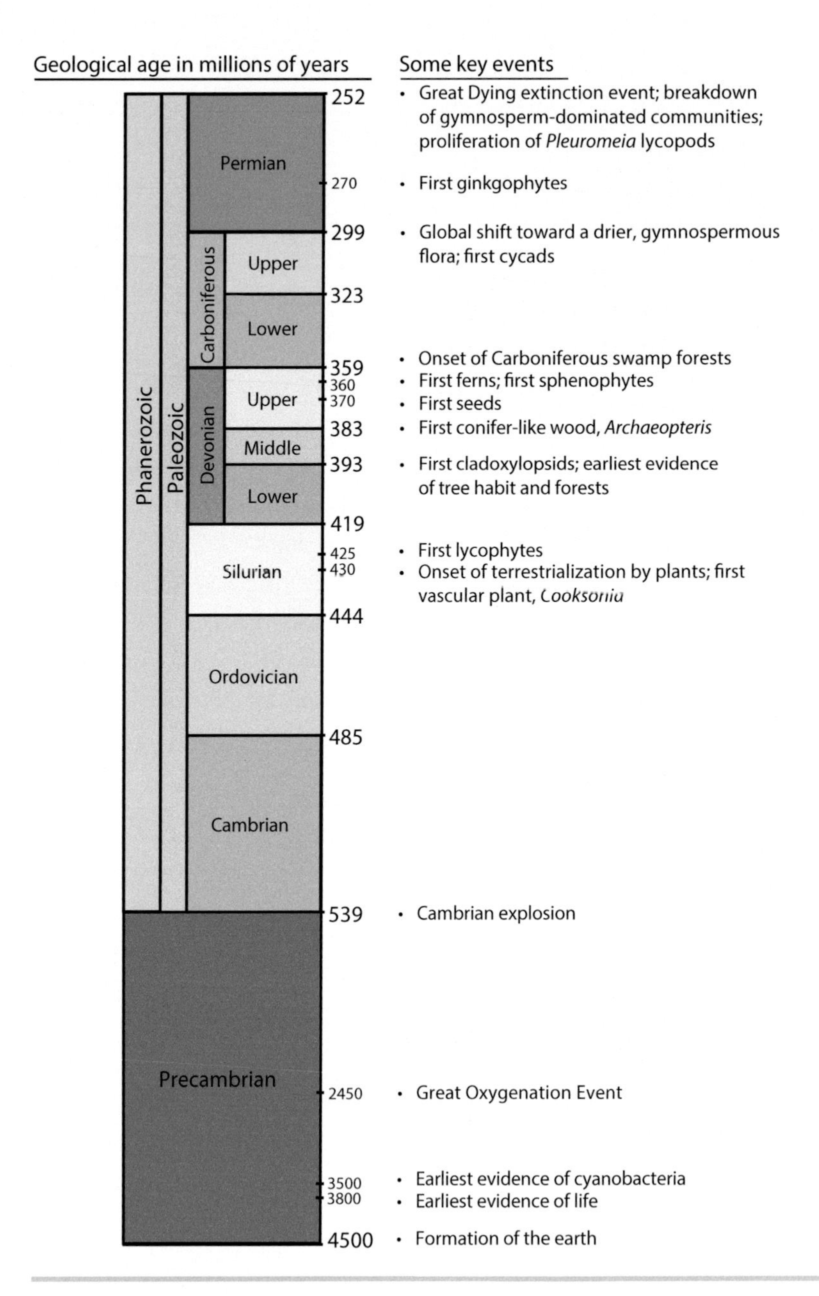

Figure 1.4. Geological time chart with a timeline of key paleontological events discussed in this book. The time chart begins at the base of the Precambrian, with the formation of the earth, extends to the end-Permian, then resumes with the Lower

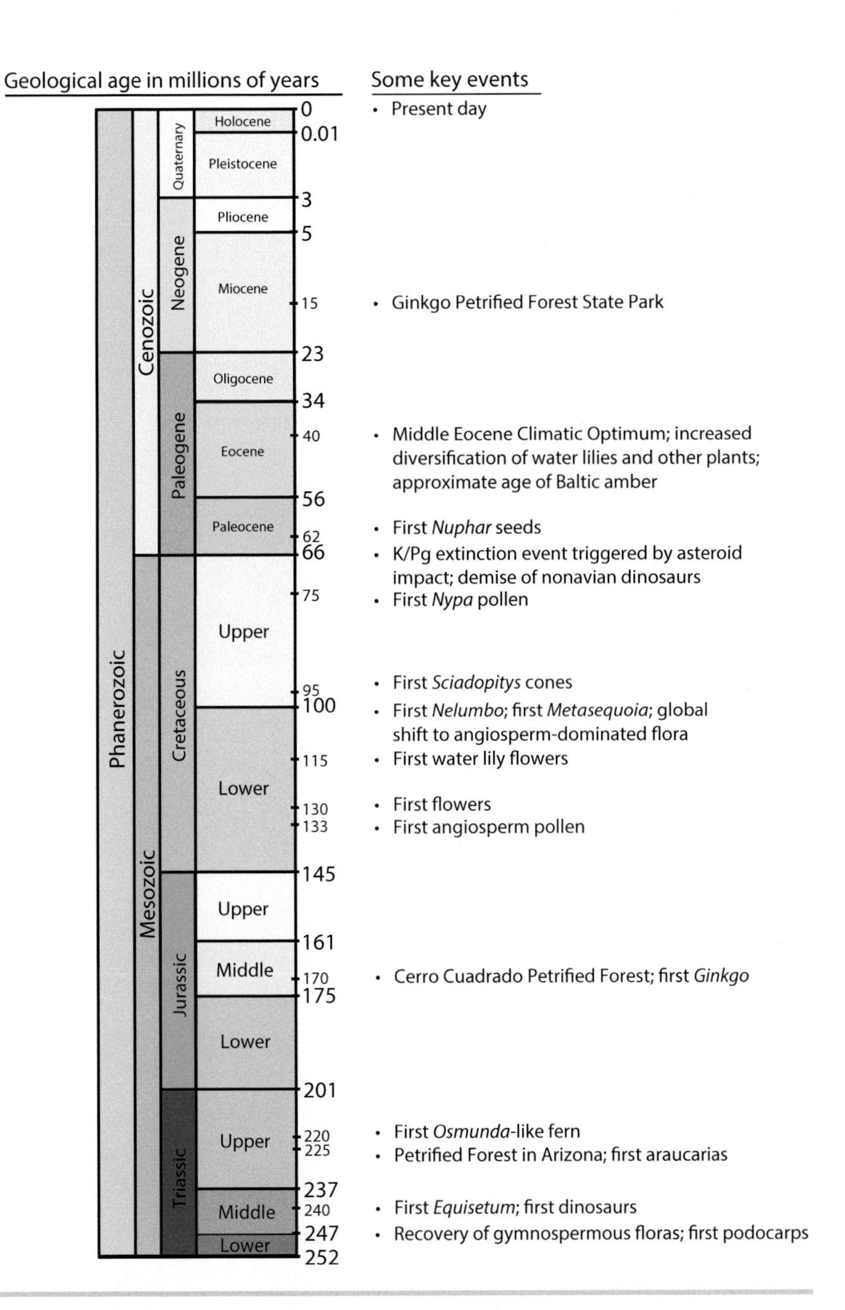

Triassic and continues to the present day. Geological ages are rounded to the nearest whole number. Based on the International Chronostratigraphic Chart, version 9/2023, compiled by the International Commission on Stratigraphy.

Figure 1.5. The last rays of the setting sun on stromatolites and Australian pied cormorants at Shark Bay, Western Australia. Photo by author.

global mangrove vegetation consisted in great part of *Nypa* palms. In contrast, in the present-day world *Nypa* palms mostly survive on outer fringes of mangrove environments along shorelines of the eastern Indian and western Pacific Oceans, having been vastly outcompeted by other, more robust mangrove species.

Among the many plants discussed in this book, two genera were initially found in the fossil record, thought extinct, then later discovered alive. Like the coelacanth *Latimeria*, they are referred to as "Lazarus taxa," alluding to the New Testament story in which Jesus Christ raises Lazarus from the dead. As with humans, finding a plant genus alive and well after it had been declared extinct seems miraculous. Instead of divine intervention, though, the discovery and recognition of the dawn redwood, *Metasequoia*, in China in the 1940s and of the Wollemi pine, *Wollemia*, in Australia a half century later can be attributed to the plant adventurers who were stubbornly persistent in their quest to identify trees unknown to them. Also deserving of recognition are the botanists who did the scholarly follow-up work of collecting and documenting more material as well as describing, classifying, and validating the identity of these trees (fig. 1.6).

As tenacious survivors and relicts of ancient lineages, several plants in this book are essentially taxonomic loners, present in small numbers today compared with their greater biodiversity in the past. These

Figure 1.6. *Left to right*: botanists C. T. Hwa, W. C. Cheng, and K. L. Chu on the trail of *Metasequoia* as a living fossil, standing at the base of the then largest tree of *M. glyptostroboides*, near Wangkia-ying (now Wangying Town in Lichuan City), Hubei Province, in August 1948. Cheng was one of the two authors who first published that *Metasequoia* was alive, not extinct. He visited the living population of *Metasequoia* in central China to study its ecology just three months after the scientific announcement of this new world sensation in May 1948. Photo © President and Fellows of Harvard College. Arnold Arboretum Archives, Boston.

include genera from throughout the plant kingdom, from the free-sporing plant *Equisetum*, to seed-bearing gymnosperms such as *Ginkgo*, *Metasequoia*, *Sciadopitys*, and *Wollemia*, to an angiosperm like *Nypa*.

As intriguing as living fossils may be, the story of fossil plants involves more than the dwindling of lineages. Understanding plants in the fossil record and their evolution is about elucidating the key evolutionary innovations that plants have developed to surmount the physical challenges of their environment and to cope with other organisms. Extremely early in the history of life on earth, for example, they learned to use carbon dioxide, water, and sunshine to make bigger and better plant bodies—photosynthesis. The oxygen released during this process accumulated in our atmosphere over eons and made it possible for oxygen-breathing organisms to appear on earth. The bigger, more complex plant bodies led to herbivory by nonphotosynthetic organisms (animals), positioning plants as the primary producers at the base of nearly every food chain in the world.

Plants then made their way out of the oceans to conquer land and in doing so had to develop more intricate bodies with adaptive morphological structures. These adaptations helped the early colonizing plants to survive the conditions of a harsh atmosphere, as well as to grow taller and compete with neighboring plants for more light. Modifications in the plant body affected the cuticle, stomata, primitive roots, water- and food-conducting systems, and stems. Land plants learned how to wrap their precious embryos into waterproof seed packages and place their seeds into protective woody cones. Plants also had to learn to defend themselves from herbivores morphologically and chemically. They even went as far as to turn the tables on plant-animal interactions by playing off of the attraction of insects to plants. One of plants' greatest achievements was to devise various colorful and scentful ways to persuade mobile organisms, such as insects, birds, and mammals, to pick up their genetic material from flowers and spread it among members of the same plant species for cross-pollination and, hence, increase the genetic fitness of the species.

To get a better sense of some of the key innovations developed by plants, chapters on various aspects of plant evolution—different strategies on how to grow tall, the seed and woody cone habits, and the advent of flowers—are interspersed with the chapters on living fossils. For plant lovers, these are all exciting historical processes that occurred in the evolution of plants. However, perhaps we should consider them from the perspective of nonbotanists in the more poetic words of authors, playwrights, and philosophers:

> Fall. Stand. Learn. Adapt.
>
> —Mike Norton, in *Fighting for Redemption*

> From a small seed a mighty trunk may grow.
>
> —Aeschylus, in *Choepori*

> The earth laughs in flowers.
>
> —Ralph Waldo Emerson, in *Hamatreya*

FURTHER READING

Carus, C. G. 1818. Beitrag zur Lehre von den lebendigen Fossilien. *Schriften der Dresdner Gesellschaft für Mineralogie* 1:135–45.

Darwin, C. 1859. *On the Origin of Species by Means of Natural Selection, or The Preservation of Favoured Races in the Struggle for Life*. London: John Murray.

Lidgard, S., and A. C. Love. 2018. Rethinking living fossils. *BioScience* 68:760–70.

Schopf, T. J. M. 1984. Rates of evolution and the notion of "living fossils." *Annual Review of Earth and Planetary Sciences* 12:245–92.

Turner, D., and J. Han. 2023. Living fossils and conservation values. *Frontiers in Earth Science* 11:1086066.

Stromatolites at Shark Bay, Australia

CHAPTER 2

Cyanobacteria and Stromatolites

The Toughest and Longest-Lived Green Survivors

In the beginning there was light. In fact, there was plenty of sunlight on earth some 4.5 billion years ago. However, that primeval atmosphere was very different from the atmosphere today. There was no free oxygen, or very little, in the air, nor was there a protective ozone layer. Instead, the atmosphere was composed of water vapor and gaseous carbon dioxide and sulfur dioxide, which organisms of today's world would find toxic. Hence, early earth's first life forms had to evolve to metabolize sulfur, not oxygen. In fact, what we need for life now—an oxygenated atmosphere—would have been deadly for them.

These first life forms were single-celled, anaerobic microbes. They did not have nuclei, as many cells do today, but were unicellular bacteria, also known as prokaryotes. Arising from these primordial organisms, a special type of photosynthesizing bacteria, called *cyanobacteria*, evolved membranes and pigments within their cells that could capture light energy from the sun and convert it into chemical energy that they could use for their metabolism. All it took was sunlight, carbon dioxide, and water, which were readily present in the early atmosphere.

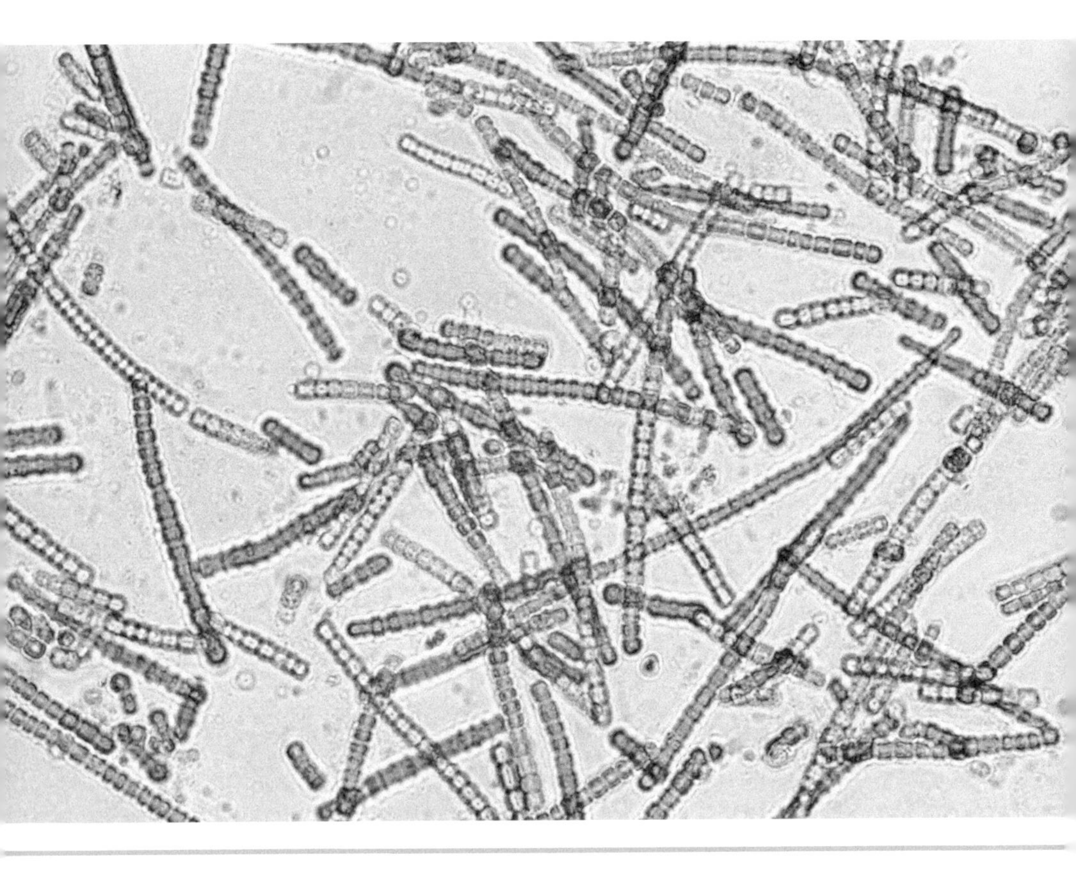

Figure 2.1. Solitary filaments of the cyanobacterium *Cylindrospermum*, photographed with a compound microscope. Note the blue-green color that is so characteristic of the cyanobacteria. Photo by Willem van Aken, CSIRO ScienceImage 4203, Creative Commons Attribution-ShareAlike 3.0 Unported license. © Copyright CSIRO Australia.

The cyanobacteria became the very first organisms that could photosynthesize, converting sunlight into stored energy (fig. 2.1). This happened about 3.5 billion years ago, one billion years after the formation of our planet.

The end products of photosynthesis are glucose, which is used by living organisms as a major source of energy and for the production of new cells, and oxygen, which is in this case a waste product. The free oxygen produced by the cyanobacteria on early earth was first absorbed by dissolved iron or other reducing minerals in rocks, resulting in the deep red, banded iron formations in Minnesota and Western Australia (fig. 2.2). These rocks with red and gray layers are considered an important line of evidence in documenting the history of life.

Approximately 2,450 million years ago, when the chemical capacity of the minerals to capture and bind to free oxygen was filled, gaseous oxygen began to accumulate in the atmosphere. Thus began the Great Oxygenation Event, marking a tremendous transformation in the composition of the atmosphere. The sulfur-metabolizing bacteria

Figure 2.2. The polished surface of a rock from a banded iron formation in Australia, showing the deep red bands resulting from cyanobacteria-generated oxygen in the atmosphere. The rock measures 53.6 cm (21 inches) across at its widest point. Photo by Georg Oleschinski.

that had evolved under oxygen-free conditions would have suffered irreparably in the oxygenated conditions. However, an atmosphere with free oxygen opened the way for the biological diversification and proliferation of organisms that could use this atmospheric gas. This included, for example, free-living mitochondria. With the passing of evolutionary time, the once-independent mitochondria became endosymbionts inside eukaryotic cells and began supplying energy to their hosts through cellular respiration with oxygen.

As the oxygen level in the atmosphere slowly built up, it prompted the evolution of more complex, multicellular organisms. Also arising from the free oxygen was an ozone layer, which helped to shield organisms from harmful UV radiation from the sun. Hence, the closing of one door for the sulfur bacteria led to the opening of another for the evolution of the diversity of life on earth today.

Through it all, the cyanobacteria survived. There is hard evidence of cyanobacteria starting from the first appearance of cells near the start of the Precambrian and continuing throughout this long geological eon. Fossil evidence of cyanobacteria from 3.5 billion to 2.5 billion years ago can be found in ancient rocks at nearly 50 different sites around the world. As the first organisms on earth, their morphologies are simple: single cells, solitary filaments, or layered cyanobacterial colonies. These simple morphologies have persisted until today and are thus used to identify cyanobacteria in the fossil record.

The layered colonies are the most striking and most easily recognized fossils of cyanobacteria (fig. 2.3). The colonies may take various forms, from flat-lying mats to dome-topped mounds to columnar structures. When mound-shaped, these fossils are called *stromatolites*. Common to each of these forms are a series of thin, millimeter-scale, light and dark bands made up of alternating layers of carbonate minerals and entombed organic material.

The layered, or laminated, mounds of stromatolites are the most striking and distinctive types of cyanobacterial fossils. The size of the mounds found in the Precambrian fossil record can vary from a few centimeters to 10 m (33 ft) in diameter. While it is usually thought

Figure 2.3. Stromatolites from the Neoproterozoic Weiji Formation, between 800 million and 1 billion million years old, at Xuzhou, northwest of Nanjing, China. *Left*, an oblique section through a colony of narrow stromatolites, showing their upward layered growth and columnar shapes. *Right*, a close-up view of a cut through a few stromatolites from the same locality, illustrating their concentric layers of microbial growth. Photos by author.

that filamentous cyanobacteria were responsible for producing them, single-celled cyanobacteria and green algae may also have played a role. Sometimes the cyanobacterial layers grow around a central core, such as a bit of shell, producing a spherical structure called an *oncolite*. When oncolites are cut open with a rock saw, concentric rings of light and dark bands show how the cyanobacterial mats grew around a nucleus to form the ball-shaped fossil. Oncolites are thought to form in swash zones, where waves constantly roll the cyanobacterial colonies around and shape them into rounded shapes. Oncolites are usually smaller than the size of a fist, reaching a maximum of 15 cm (6 in) in diameter.

The laminations in stromatolites represent the continual, repeating cycles of the microbial colony fighting to photosynthesize, grow, and survive. The dark, organic-rich layers are produced from the growth of cyanobacteria toward the sun to capture light for photosynthesis. On their surface, the cyanobacteria and other microbes have a sticky mucilaginous or gelatinous layer that binds to fine particles of calcium carbonate. Or they might precipitate carbonate directly from seawater. The carbonate gets bound to the sticky cyanobacteria and forms a horizontal layer over it. In any case, when the carbonate precipitates in seawater, the carbonate layer can start to harden. Before this happens, though, to keep from being trapped under the hard layer of minerals, the photosynthesizing filaments of a cyanobacterial colony must grow upward and through this initial carbonate layer to reach for the sun. Thus, little by little, layers of organic matter and carbonate build up and harden, producing a mound or tower-like colony made of limestone.

Over time, the growing mounds of carbonate can become taller columns with living cyanobacteria only at the top. Sometimes the colonies overtop themselves, forming a sort of rounded mushroom cap when viewed from the side. As sun worshippers, the upper layer of living cyanobacteria cannot be covered too deeply by water; that is why the mounds or columns are found in shallow coastal waters or along the shorelines of lakes.

In fact, throughout all paleontological history, stromatolites have been restricted to clear, shallow zones where sunlight can easily penetrate water. However, the sudden proliferation of marine invertebrates during the Cambrian explosion, 539 million years ago, posed a challenge to the colonial cyanobacteria with the rise of marine herbivores. Stromatolites now faced a new restriction to their exuberant growth and global expansion. Their living filaments were now being checked by the grazing of marine mollusks such as chitons and primitive sea snails. While all marine plants, such as green algae, kelp, and seagrass, suffer from the effects of herbivory, stromatolites managed to get around the problem by expanding into areas with hypersaline water. The supersalty water is too harsh for the marine herbivores and thus keeps them from devouring the cyanobacterial filaments.

Despite their general confinement to hypersaline habitats, stromatolites not only survived but also thrived for hundreds of millions of years. In the Upper Triassic, 201 million years ago, for example, stro-

Figure 2.4. Two hundred-million-year-old stromatolites from Upper Triassic Cotham Marble. *Top left*, slope of an inactive quarry on a farm in the idyllic English countryside, near Bristol. *Top right*, university students at the top of the slope searching for the discus-shaped specimens. *Bottom*, several slices through a stromatolite, featuring a central section with a classic view known in Victorian times as Landscape Marble, with full-crowned trees in between hedges and growing on soil horizons. Maximum width is 16.5 cm (6.5 in). Photos by author.

matolites colonized the shallow waters in what is today a peaceful, pastoral area of southwestern England. Here, on a farm near Bristol, disk-shaped lumps on the slopes of an old rock quarry are collected by university students on paleontology field trips. When the limestone rocks are sawed into a series of slabs, the structure of the cyanobacterial colonies becomes apparent (fig. 2.4). Curiously, the upward growth of the cyanobacteria and natural coloration processes during fossilization have created miniature sylvan scenes on the cut surfaces of the rock slabs.

Most cut surfaces of the rock slabs reveal thin layers across the lower part of the rock, resembling soil horizons. Rooted on these layers of soil is a wooded landscape of mature trees with full, orange-colored crowns and gray-colored trunks. Between the trees can be seen low-growing, black hedges and a layered, light-colored sky. In Victorian times, highly polished slabs of Cotham Marble, sometimes called Landscape Marble, depicting such whimsical scenes were popular ornamental objects of wonder, appealing to the romantic sentiment and love of curios in the late nineteenth century.

Today, there are several places on earth where living stromatolites thrive. As one would expect, their need for supersalty water means that they are usually found in arid areas, where evaporation keeps the salt content of the water high and the grazing herbivores out. The best-known site with living stromatolites is Shark Bay in Western Australia (fig. 2.5). Shark Bay is located about 800 km (500 mi) north of Perth, at the westernmost point of the continent. Declared a UNESCO World Heritage site in 1991, Shark Bay is a hypersaline lagoon with a spectacular meandering coastline, the world's largest and biologically most diverse seagrass meadows, and a large community of dugongs, large aquatic mammals similar to manatees but native to the Old World. However, it is the colonies of stromatolites at the Hamelin Pool Marine Nature Reserve that make Shark Bay so remarkable.

At Shark Bay, the shallowness of the lagoonal basin and the hot, dry climate drive up the salinity of the marine water, which is approximately double the salinity of normal seawater. Thus, the cyanobacte-

Figure 2.5. Stromatolites at low tide in the hypersaline embayment at the Hamelin Pool Marine Nature Reserve, Shark Bay, Western Australia, in 2019. Photo by author.

rial mounds are protected from marine herbivores by the hypersaline conditions. Untroubled by grazing, this extensive colony of stromatolites is thought to have been established one thousand years ago in the southern part of the lagoon.

Stromatolites will also grow in hypersaline inland lakes, where evaporation is high, freshwater input is low, and grazing invertebrates are scarce. Although the salt content of these waters may be very high, these are not marine environments but belong to the freshwater, continental realm (fig. 2.6). Here, the cyanobacterial colonies are sometimes called freshwater microbialites. They commonly consist not only

Figure 2.6. Rounded mounds of microbialites along the shoreline of Lake Clifton, an inland lake south of Perth in Western Australia. Photo by author.

of cyanobacteria but also of algae and other microbes and form low, rounded mounds or even grow into massive towers of layers of limestone. While there are inland lakes containing microbialites all over the world, one can see them along the shoreline of Antelope Island, in the Great Salt Lake in Utah (fig. 2.7), and in Mono Lake in eastern California (fig. 2.8).

In the botanical world, organisms are judged on the basis of their pigmentation. For photosynthesis, chlorophyll is the most important pigment and comes in seven different molecular variations—called chlorophyll *a, b,* c_1, c_2, c_3, *d, f*—which absorb light in slightly different parts of the wavelength spectrum. While chlorophyll *a* is universal and chlorophyll *b* is found in most plants, cyanobacteria have the additional pigments chlorophyll *d* and *f*. Perhaps the Australian stromatolites have a special pigmentary advantage, for chlorophyll *f* was recently discovered in the Shark Bay cyanobacteria. This variant absorbs light waves farther into the red part of the light spectrum than other

Figure 2.7. Low-mounded microbialites in the warm, shallow waters around Antelope Island in the Great Salt Lake, Utah. Photo by Brooks Britt.

Figure 2.8. The microbe-precipitated tufa limestones at Mono Lake, a saline alkaline lake in Mono County, California, where the towers of calcium carbonate sometimes form tall, fanciful spires. Photo by author.

Figure 2.9. The distinctive dark blue-green color of the cyanobacteria of Shark Bay, Western Australia, cultivated here in an aquarium in the museum at Telegraph Station Hamelin. The cyanobacteria produce oxygen, which is being released here as gaseous bubbles. Photo by Louise Woo.

chlorophylls and may be more beneficial for photosynthesis in environments with higher levels of infrared light and lower levels of ultraviolet light.

Also important are accessory pigments, which determine plants' color and identity. In the case of the cyanobacteria, an accessory pigment called phycobilin imparts a bluish tinge. The combination of green and blue pigments gives cyanobacteria their very characteristic color. In fact, until relatively recently this group was known as blue-green algae because of its green and blue pigmentation (fig. 2.9). Now recognized as bacteria, the group's new name, cyanobacteria, emphasizes its coloration, as *cyan* is the name of its distinctive greenish blue color and comes from the Greek word *kuaneos*, meaning "dark blue."

Although science continues to discover more and more about the botanical affinities, growth, physiology, and pigmentation of the cyanobacteria, there is no doubt that they were the major players in the establishment of a habitable, oxygenated earth 3.5 billion years ago. They have survived since the beginnings of life on earth, shifting habitats and adapting to the new challenges constantly thrown at them by environmental changes and biotic evolution. For this reason, the cyanobacteria and the structures that they form, stromatolites, are truly the toughest and longest-lived green survivors on earth.

FURTHER READING

Knoll, A. H. 2004. *Life on a Young Planet: The First Three Billion Years of Evolution on Earth.* Princeton, NJ: Princeton University Press.

Parks and Wildlife Service. n.d. Stromatolites. https://www.sharkbay.org/place/hamelin-pool/stromatolites/.

Scalice, D. 2011. Earth's early atmosphere: An update. https://astrobiology.nasa.gov/news/earths-early-atmosphere-an-update/.

Schopf, J. W. 2012. The fossil record of cyanobacteria. In *Ecology of Cyanobacteria II: Their Diversity in Space and Time*, ed. B. A. Whitton, 15–36. Dordrecht: Springer.

PART II

CHAPTER 3

Standing Tall

Terrestrialization, the colonization of land by marine organisms, was a major event in the history of life. For plants living in the ocean, the move onto land was fraught with new physiological challenges requiring morphological and anatomical solutions. To keep from drying out, plants developed a waxy layer on their stems and leaves called cuticle. To absorb water from the soil, plants grew rhizoids and roots. To draw water up their stems, plants evolved vascular tissue. To breathe, plants formed little holes in their leaves and shoots called stomata, which were opened and closed by kidney-shaped guard cells. And to stand tall, plants engaged in a series of creative experiments that would give them the support and strength to stand upright and grow toward the sun.

The earliest vascular plant known to date is *Cooksonia*, which appeared in the Silurian, 430 million years ago. *Cooksonia* was a tiny, slender plant only a few centimeters (less than an inch) high and only up to 3 mm (0.1 in) wide. In some flattened fossil shoots, a vertical stripe of darker tissue running down the center is interpreted as vascular tissue. This vascular tissue was used to conduct water from the soil by the little root-like structures called rhizoids to the top of the plant. The vascular tissue was composed of cells with thickened walls, which not only lent rigidity to the cells but also gave mechanical strength to the stem. *Cooksonia* did not reach any great heights in a literal sense, but this slender little stem was the first land plant to stand tall.

The genus name *Cooksonia* is a tribute to the nineteenth-century Australian botanist Isabel Cookson. Cookson specialized in fossil plants and pollen and was interested in early vascular plants from the Silurian and the Devonian, as well as in early land plant evolution. Her research collaboration with William Henry Lang at the University of Manchester in England led Lang to bestow the ultimate honor on his esteemed colleague by naming the first vascular plant after her.

The First Forests

Huge changes in plant height occurred in the Devonian. The tree habit first shows up in the Middle Devonian, about 40 million years after *Cooksonia* conquered land. Some of the geologically oldest tree-like plants were discovered in the small town of Lindlar, some 36 km east of Cologne, Germany. These arborescent plants were at least 2 m (6.5 ft) high, though they may have attained heights of 3–4 m (10–13 ft), and grew in multigeneration forests (fig. 3.1). They had a slender trunk that flared near the roots and bore nubby branch bases along the top part of the trunk. At the very top of the trunk, each plant had a wide, rounded crown of feathery branches. *Calamophyton*, the genus name of this Devonian tree-like plant, comes from the Greek *kálamos*, meaning "pen" or "quill," and the Greek *phytón*, meaning "plant."

Although *Calamophyton*, with its slender trunk with feathery fronds on top, may have superficially resembled a tree fern, it differed from tree ferns of later paleontological time. The trunk of *Calamophyton* stood upright on the basis of its own inner strength. In contrast, the Carboniferous tree fern *Psaronius* supports its trunk with an outer

Figure 3.1. Reconstruction by Peter Giesen of a Middle Devonian forest with *Calamophyton primaevum* from western Germany. This 390-million-year-old fern was the geologically oldest or among the oldest tree-like plants on earth and stood up to 2 m tall. Courtesy of Peter Giesen.

mantle of roots (see fig. 3.3), much as modern tree ferns such as *Cyathea* do today (see figs. 6.5 and 8.3, right). The enigmatic, forest-forming, 390-million-year-old *Calamophyton* is placed in an extinct class of plants known as the Cladoxylopsida and thought to be related to ferns and horsetails.

Across the ocean, at a slightly younger site in the small town of Gilboa, New York, a quarry with hundreds of tree stumps in original growth position indicates that a dense forest of tree-like plants also flourished there. This 387-million-year-old stump forest was discovered during the excavation of sandstone to build the nearby Gilboa Dam in the early 1920s. Pioneering work on the fossil stumps was conducted by Winifred Goldring, the first female New York State paleontologist. In 1924, Goldring created a diorama that depicted the Gilboa Fossil Forest, one of the first reconstructions of a Devonian terrestrial habitat. After the extraction of building stone was completed in the Riverside Quarry, the fossil forest site was filled in, sealing off access. Nevertheless, the Gilboa Fossil Forest had become world famous, and until the discovery of the Lindlar forest in Germany it was renowned among paleobotanists as the first forest in the fossil record.

Eighty-six years later, in 2010, the Gilboa quarry was exposed for a short time to retrieve more building stone. American paleobotanists rushed in to study the details of the stumps, roots, and soil horizon there, collecting data for only 10 days for a new interpretation of the Gilboa Fossil Forest as a complex ecosystem. What the paleobotanists found was that the tree stumps of the Gilboa Fossil Forest are large and measure up to 1 m (39 in) in diameter. Because the trees at this site are preserved mainly as sandstone casts, there is little evidence of their internal anatomy. However, fossil foliage with spores shows that these arborescent plants pertained to the class Cladoxylopsida, the same group of tree-like plants found at Lindlar, Germany. The Gilboa trees, which were named *Eospermatopteris*, bore foliage in a tuft at the top of a laterally leafless trunk.

The 387-million-year-old site at Gilboa represents a mixed-aged forest of mostly *Eospermatopteris* trees that stood up to 8 m (26 ft) tall. This is a remarkable height for plants so early in evolutionary history. Yet, because there was no internal anatomy preserved in these fossil stumps, how these ancient ferns were able to attain such heights remained a mystery until the next discovery was made in the Devonian of China.

At a slightly younger site in Xinjiang Province, in northwestern China, anatomically preserved cladoxylopsid trees in the Upper Devonian revealed the secret behind their ability to grow tall. In their trunks, soft tissue was found to be stiffened with two concentric cylinders of strands of water-conducting tissue running through the core area of the trunk. The vascular strands, which had cells with thick, reinforced cell walls, divided and came together again on their way through the tree to form a net-like mesh of supporting tissue in the outer part of the trunk. Here, the network of vascular strands also gave rise to roots that emerged from the tree's periphery and grew downwards along the trunk to anchor and stabilize the trunk.

Attaining height with a mostly soft-tissue trunk, which is reinforced either internally with bundles of vascular tissue or externally with a thick outer mantle of roots, is a strategy that arborescent plants with fern-like foliage have pursued from the Upper Devonian until today (see chapter 6). The Devonian Cladoxylopsida trees show that the urge to grow tall occurred early on in plant evolution. Evolving a more robust internal structure was one creative solution that allowed them to achieve great heights nearly 400 million years ago.

Why Grow Tall?

Why invest the energy and evolutionary effort to elongate vertically? Plants are driven to achieve height by their need to outcompete neighboring plants in the race to absorb more sunlight for the essential process of photosynthesis. Lifting the more sensitive tissues in leaves and reproductive organs higher above the ground also reduces herbivory by lower-dwelling insects. There may also be a protective advantage to raising delicate plant organs out of reach of ground fires. Mainly, though, it is the fight for spatial resources that drives the competition in plant height.

Another group of plants, the lycophytes, known today as clubmosses (chapter 4), also made a bid to attain a tree-like habit in the Upper Devonian. In Svalbard, Norway, 380-million-year-old lycophytes are preserved as trunk casts rooted in original growth position. These lycophytes, which were given the name *Protolepidodendropsis*, achieved a height of 4 m (13 ft), just half the height of the Devonian cladoxylopsid ferns. However, the true achievements of the lycophytes would come about a bit later in paleontological time, when they became the giants of the Carboniferous forests.

A third strategy for growing tall in the Upper Devonian was pursued by an extinct group of plants called the progymnosperms. The progymnosperm *Archaeopteris* was the first plant with massive amounts of woody tissue, which allowed the stately tree to reach a height of 30 m (nearly 100 ft) in Devonian to Carboniferous times. The fossil wood of *Archaeopteris* was similar to that of conifers today in being composed of a dense tissue with a single type of conducting cell. In fact, when the wood of the *Archaeopteris* tree was first described, it was so attractive that it was given the name *Callixylon*, derived from the Greek *kalos*, meaning "beautiful," and *xylon*, meaning "wood."

Archaeopteris also had an extensive branching system with fern-like foliage and spore-bearing reproductive structures, not seeds. This odd suite of characters in these plants—massive coniferous wood with spores—represents the transition of characteristics from free sporing to seed bearing, as well as the next key innovation in plants, seeds (chapter 7).

The Mighty Carboniferous Swamp Forests

The global swamp forests that flourished during the Carboniferous, about 359 million to 299 million years ago, were made up of a variety of plant groups. Each plant group had its own morphological strategy for outgrowing its neighbors, but some were more successful than others. It was during the Carboniferous that the lycophytes reached their acme in species diversity and in morphological stature, thus forming the framework of the Carboniferous swamp forests (fig. 3.2).

Lycophytes were by far the tallest trees in the Carboniferous swamp forests. The giant lycophyte *Lepidodendron* achieved towering

Figure 3.2. Diorama of a Carboniferous swamp forest with various giant lycophyte trees, delicate ferns, and low-growing sphenophytes that lived 310 million to 305 million years ago, on display at the Smithsonian National Museum of Natural History in Washington, DC. Photo by author.

heights of 50 m (164 ft), followed by its cousin *Sigillaria,* which, at 25 m (82 ft), was still quite sizeable. Like the Devonian Cladoxylopsida, the Carboniferous lycophytes had a core of mostly soft tissue, but they supported and stabilized their trunks externally, not internally, with incredible amounts of bark. The ratio of bark to wood in the Carboniferous lycophytes was normally 8 to 1 but sometimes as high as 20 to 1. (By comparison, the biomass bark-to-wood ratios in today's trees have been measured as 1:10 to 1:5.) While the massively thick bark of the Carboniferous lycophytes may have protected individual trees from fire, it is certain that it contributed in large part to the structural stability of these giant trees, enabling them to stand very tall.

New groups of plants, such as tree ferns, evolved during Carboniferous times. They, as well as a second, now extinct group of plants called seed ferns, formed the mid-canopy of the Carboniferous forests. The best-known Paleozoic tree fern is *Psaronius*, a member of the family Marattiaceae, which still thrives today in the tropics, for example, as the genera *Marattia* and *Angiopteris*. *Psaronius* is usually depicted as having the usual "prehistoric" look of a tree fern, with a flush of fronds at its crown and a root mantle of adventitious roots twining their way down the trunk. Its height of 10 m (nearly 33 ft) was made possible by the structural support of the root mantle, as well as by internal vascular tissue (fig. 3.3).

Another plant group growing in the Carboniferous swamps was that of the sphenophytes, a group derived from Cladoxylales, which include today the horsetail, *Equisetum* (chapter 5). The best-known Carboniferous sphenophyte is *Calamites*, which reached heights of up to 20 m (over 65 ft), making it considerably taller than its smaller, more gracile cousin *Equisetum*, which tops out today at around 1.5 m (5 ft) at temperate latitudes, although some species can reach 7.3 m (24 ft) in subtropical and tropical climes. Carboniferous *Calamites* was able to achieve its height by developing a woody stem, although the amount of wood was not as massive as in the Carboniferous gymnosperms, such as *Cordaites*.

Figure 3.3. A cross section through the silicified trunk of *Psaronius infarctus* from the Permian of Saxony, Germany. The sinuous structures in the center of the stem are parts of the vascular system, and the smaller circles in the stem periphery are the adventitious roots in the thick root mantle surrounding the tree fern trunk. Photo by Georg Oleschinski.

Along with the lycophytes, *Cordaites* formed the canopy of the Carboniferous forests. *Cordaites* could reach 1 m (3.3 ft) in diameter and 33 m (108 ft) in height. Most of its trunk consisted of dense, conifer-like wood, which allowed these gymnosperms to grow straight and tall. Some cordaites may also have had woody prop roots similar to the those of the mangrove *Rhizophora* (see fig. 17.1), which would have aided in anchoring and stabilizing the tall, heavy trees growing on swampy substrates.

These various body-building strategies created a morphologically and biologically diverse flora during Carboniferous times. However, it turns out that it was the production of massive coniferous wood that survived as the best evolutionary innovation for attaining height. Most trees today, whether ginkgoes, conifers, or angiosperms, have trunks with this type of dense, coniferous wood.

Stronger Together, with Silica

One last approach to standing upright without investing in woody tissue is carried out today by the humble horsetail *Equisetum* (chapter 5).

Figure 3.4. The *Equisetum arvense* shoots in this dense colony lean on one other to stand upright. Photo by author.

Although its cousin *Calamites* went down the road toward wood production in the Carboniferous, today's *Equisetum* is fully herbaceous, containing no wood at all. Instead, nearly all parts of the plant, especially the stem surface, are covered or filled with silica (see fig. 5.2, left). Silica produces the mechanical strength of the individual cells, which is then translated into stiffer tissues and an upright stance in the hollow shoots.

Equisetum forms thick colonies, so that the hollow, cylindrical shoots grow clustered close together. This suggests the comparison of the collective strength of a handful of straws to the weaker state of a single straw (fig. 3.4). One of the tallest horsetail species on earth today, *Equisetum giganteum*, which grows up to 5 m (16.5 ft) tall in Chile, maintains its upright stance in individual shoots by relying on the synergistic support of adjacent shoots to prevent buckling. While it is considered only partly self-supporting, *Equisetum* still stands tall.

FURTHER READING

Gastaldo, R. A., M. Bamford, J. Calder, W. A. DiMichele, R. Iannuzzi, A. Jasper, H. Kerp, S. McLoughlin, S. Opluštil, H. W. Pfefferkorn, R. Rößler, and J. Wang. 2020. The coal farms of the Late Paleozoic. In *Nature through Time*, ed. E. Martinetto, E. Tschopp, and R. A. Gastaldo, 317–43. Berlin: Springer Verlag.

———. 2020. The non-analog vegetation of the Late Paleozoic icehouse–hothouse and their coal-forming forested environments. In *Nature through Time*, ed. E. Martinetto, E. Tschopp, and R. A. Gastaldo, 291–316. Berlin: Springer Verlag.

Hong-He Xua, H.-K., C. M. Berry, W. E. Stein, Y. Wanga, P. Tanga, and Q. Fua. 2017. Unique growth strategy in the Earth's first trees revealed in silicified fossil trunks from China. *Proceedings of the National Academy of Sciences of the United States of America* 114:12009–14.

McGhee, G. R., Jr. 2018. *Carboniferous Giants and Mass Extinction: The Late Paleozoic Ice Age World*. New York: Columbia University Press.

Lycopodium clavatum

CHAPTER 4

Clubmosses & Co.

Like white-tailed deer, clubmosses are gentle, peaceful woodland beings. Unlike hoofed forest mammals, however, many clubmosses are small, creeping, and leafy (fig. 4.1). They are part of a larger group of plants known as the lycophytes, which include the clubmosses, the firmosses, the spikemosses, and the quillworts. Amazingly, lycophytes have tenaciously survived the 425 million years since the upper Silurian and are considered to be the oldest surviving lineage of vascular plants on earth.

Today's lycophytes are evergreen, perennial, and nonwoody. Their stems commonly fork into two equal branches and bear short, slender leaves. The leaves project away from the shoots, giving the plant a bristly look, though the leaves may be quite soft and flexible. Under optimal growth conditions, pioneering lycophytes can establish colonies that form widespread carpets in forests. While some lycophytes are terrestrial and rooted on soil, others are epiphytic and grow attached to trees, although not in a parasitic way. Lycophytes prefer moist conditions and are commonly found in tropical or moist temperate forests. Some lycophytes, such as bogmosses, prefer wetland habitats such

Figure 4.1. *Lycopodium clavatum*, the common clubmoss or stag's horn clubmoss, in a forest in southern Germany. Photo by Bernd Haynold, Creative Commons Attribution-ShareAlike 3.0 Unported license.

as open bogs and mires, while quillworts like even wetter habitats and thrive as submerged aquatic plants. One exception is *Selaginella lepidophylla*, known as the resurrection plant or the false rose of Jericho. This particular species of spikemoss is able to survive drought in its native Chihuahuan Desert in North America. It curls up its shoots to form one big ball as it becomes totally desiccated and then unfurls when rainy conditions return.

The best-known lycophyte genus is *Lycopodium*, called the clubmoss for its slender club-shaped strobili, or cones, which carry the spores. The clubmoss is closely related to the bogmoss *Lycopodiella*, as well as to the firmoss *Huperzia*. The firmoss, when growing upright, has the superficial appearance of a sapling fir tree, hence its descriptive common name. Despite the suffix *-moss*, these plants are not mosses, a group that does not have true conducting tissues (xylem and phloem). Instead, clubmosses are true vascular plants with xylem and phloem tissues that transport water and foodstuffs within the plant body. There are also epiphytic species of *Huperzia* that grow in a pendulous manner (fig. 4.2), which are instead called tassel ferns, referring to the dangling nature of their shoots. However, *Huperzia* species are not ferns either.

There are two other major groups of lycophytes: *Selaginella*, the spikemoss, and *Isoetes*, the quillwort (fig. 4.3). These two groups differ from the clubmosses and firmosses in having spores of two different sizes, a trait called heterospory. In heterosporous plants, the female spores are larger than the male ones. In what is thought to be an early stage in the evolution of sexual differentiation in plants, the two types of spores are held in separate spore-bearing structures. One advantage of heterospory is that fewer, larger spores could each contain greater resources to support the developing seedling. In contrast, the production of numerous small spores would be more practical for easier dispersal and greater cross-fertilization.

Surprisingly, although lycophytes are quiet and inconspicuous plants, they have been used in myriad cultural and economic ways through the ages. Perhaps one of the merriest uses of leafy lycophyte shoots is in Christmas decoration. *Diphasiastrum digitatum* is a lycophyte species native to eastern North America that grows as a long, horizontal stem along the ground. From the stem arise a series of miniature "trees" with spreading, fan-like crowns that resemble cedar foliage or pine trees. Known popularly as the fan clubmoss or trailing pine, this lycophyte's foliage is shiny and dark green, making it attractive in Christmas wreaths. At one point in time, the fan clubmosses

Figure 4.2. The pendulous, forked shoots of the tassel fern *Huperzia phlegmarioides* at the Botanical Garden in Berlin-Dahlem. Photo by Krzysztof Ziarnek, Creative Commons Attribution-ShareAlike 4.0 International license.

Figure 4.3. *Left*, the spikemoss *Selaginella bryopteris* at the Royal Botanical Garden of Madrid, Spain. Photo by Luis Fernández García, Creative Commons Attribution-ShareAlike 2.1 Spain license. *Right*, the black-spored quillwort, *Isoetes melanospora*, an endangered species, growing in a shallow granite pool on top of Arabia Mountain, Georgia, in the United States. Photo by AHR12, Creative Commons Attribution-ShareAlike 3.0 Unported license.

were so widely gathered and sold as Christmas greenery that populations in the wild suffered as a result. In the meantime, however, the fan clubmoss has recovered throughout its range and is no longer in decline or endangered.

In addition to their use in decor, lycophytes have been used medicinally. Both American Indian and European cultures brewed teas from the leaves and spores. These brews are used for healing urinary-tract problems, diarrhea, and other digestive-tract problems, relieving headaches and skin ailments, and inducing labor in pregnancy. Mainly, though, it has been the spores of the clubmosses that have been economically important. Clubmoss spores, which are highly water-repellent, have proven beneficial in treating skin rashes and wounds and have also found application as a baby powder or as a coating for pills. Masses of clubmoss spores, called "lycopodium powder," can be used as a lubricating dust inside latex medical gloves or even for condoms. The most common sources of lycopodium powder are *Lycopodium clavatum* (see fig. 4.1) and *Diphasiastrum digitatum* (the Christmas greenery), both widespread species with abundant spores that are easy to gather.

Lycopodium powder itself appears as a yellow-tan fine dust. The tiny spores, each with a diameter of 33 micrometers and thus a large surface area per unit of volume, contain a high oil content that is easily set on fire. With these properties, lycopodium powder suspended in air can cause the rapid combustion of the fine particles (the spores), creating a spontaneous dust explosion. Naturally, a sudden explosion with a flash of light can also be ignited by a flame. The highly flammable nature of lycopodium powder made the spores popular among Native Americans for ceremonial purposes; medicine men would toss them into a fire to create a magical flash of light. Similarly, lycopodium powder is sometimes employed to produce other theatrical special effects, such as when a bright burst of light is desired in flash photography, stage productions, fireworks, or chemistry lab demonstrations.

In studies of fossil pollen and spores, *Lycopodium* spores play a special role in determining the abundance and diversity of ancient species preserved in rock samples. In this case, a tablet with a known number of spores of *Lycopodium clavatum* is added to a pollen sample during lab preparation, thereby "spiking" the sample. When the prepared fossil material is smeared onto glass slides and studied under a light microscope, the relative count of modern *Lycopodium* spores serves as a way to determine the absolute abundance of pollen grains and other spores in the prepared sample.

Today, the lycophytes make up a species-rich group of plants comprising 3 orders, 3 families, 18 genera, and roughly 338 species. Despite this impressive show of diversity, the group is a mere shadow of what it was in deep paleontological time. Early on in the evolution of the terrestrial biosphere, the lycophytes were dominant players on the global stage.

In the Silurian and the Devonian, plants experimented with many different forms, growth strategies, and life cycles. The early lycophytes were a major plant group that decided to take a different turn down the evolutionary road, thus splitting the course of plant diversification (fig. 4.4). In the Devonian, they were small and nonwoody and bore short, thin, leaf-like structures that spiraled around the stems. These leaf-like structures were not megaphylls such as those borne by most plants today but instead microphylls. Microphylls are simple leaves with a single vein and are commonly narrow and spine-like (fig. 4.5). Megaphylls differ from microphylls in having branching venation, which allows them to produce broader and more complex leaf shapes.

In the Carboniferous, lycophytes reigned supreme. Giant lycophyte trees dominated the swamp forests of the coal age, some 359 mil-

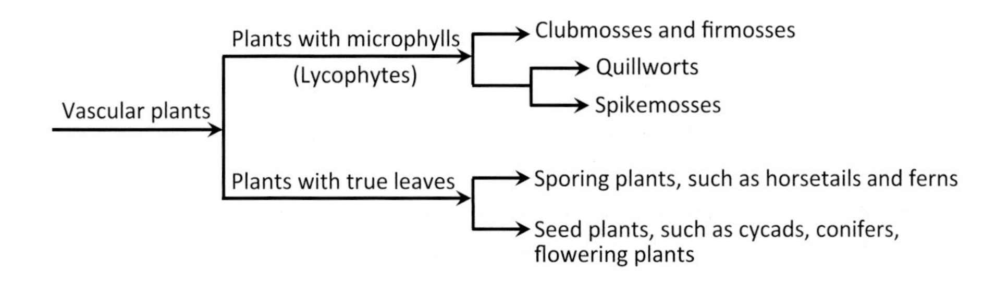

Figure 4.4. A simplified tree of plant evolution highlighting the independent path taken by the lycophytes (clubmosses and close relatives). The onset of land plant evolution begins on the left. The first fork in the road of vascular plant evolution—developing microphylls or true leaves—happened early, some 425 million years ago and led to a major split in the plant kingdom. Modified from Pteridophyte Phylogeny Group I (2016).

Figure 4.5. Devonian lycophytes. *Left*, the earliest lycophyte in the Early Devonian of the Rhineland in Germany, *Estinnophyton wahnbachense*, was small in stature and had the single-veined, spine-like microphylls so typical of the clubmoss *Lycopodium* today. Here, each microphyll measures ca. 3 mm (0.1 in). Photo by Georg Oleschinski. *Right*, in the Middle Devonian lycophytes retained their simple, slender microphylls but attained an increase in stem girth, as shown, for example, in a lycophyte found in Wisconsin called *Archaeosigillaria*. In this broader stem, the spiral arrangement of the leaves and leaf bases shows up more clearly. Photo by Kenneth Gass, Creative Commons Attribution-ShareAlike 4.0 International license.

lion to 299 million years ago. During these 60 million years, lycophytes learned to build towering, tree-like columns up to 50 m (more than 164 ft) tall. Some were massive in diameter too, stretching to 2 m (6.5 ft) at their base. To outcompete other plant groups for sunlight (see also chapter 3), the early lycophytes needed to form taller plant bodies and thus put their stock into the evolution of massive bark on the outside of their soft trunks. The thick bark of the mid-Paleozoic lycophytes gave their trunks the mechanical stability to reach unprecedented heights.

Carboniferous lycophytes' reliance on a thick layer of bark may not have been wholly due to the need to support their trunks in their desire to grow taller. Thick bark may also have served as protection from herbivorous insects. During the Carboniferous, insects reached enormous sizes. For example, *Meganeura*, related to today's dragonflies and damselflies, was the largest insect ever to exist on the earth. It had a wingspan of more than 70 cm (28 in), although this particular group of carnivorous insects was likely more interested in animal prey than in the soft tissues under the bark of lycophyte trees.

The Carboniferous lycophytes bore their leaves on the top of their trunks. The best-known genera are *Lepidodendron* and *Sigillaria*. Lepidodendrids were the tallest lycophytes, with a height of 50 m (164 ft), and had a more massive trunk and a rounded crown. *Sigillaria* was a bit smaller at 30 m (98 ft), more gracile, and commonly branched on top in two elongate, bushy tufts (fig. 4.6). As the Carboniferous lycophytes grew taller, the older branches and leaves naturally fell off the lower part of the trunk, leaving scars where the leaves had once been attached, hence the common name "scale tree" for *Lepidodendron*. Because the leaves were born in tight spirals, the exposed leaf scars around the trunk were arranged in intricate spiral patterns that are distinctive of the lycophyte genera. These patterns are so iconic that the Natural History Museum in London, which opened in 1881, was built with columns bearing *Lepidodendron* leaf scars as a tribute to the aptly named scale trees, which flourished in the Carboniferous swamp forests in what is today Britain (fig. 4.7).

Figure 4.6. Common lycophytes in the swamp forests of the Carboniferous were *Lepidodendron* (*left*) and *Sigillaria* (*right*), which differed not only in their maximum height but also in the girth of their trunk, in the branching at the top, as well as in how they bore their leaves and dangling cones. Reconstructions by Tim Bertelink, Creative Commons Attribution-ShareAlike 4.0 International license.

As the global flora shifted from arborescent lycophytes in the Carboniferous to a great number of gymnosperms, such as *Ginkgo* and conifers with true wood, in the Permian, owing to the drying up of coastal swamps and global climate change, the abundance and physical stature of the lycophytes began to shrink. Yet, an interesting phenomenon occurred after the end-Permian, 252 million years ago, fol-

Figure 4.7. The Natural History Museum building in London is graced with iconic motifs of the scale trees of the Carboniferous. *Left,* a column bearing the leaf scars of *Lepidodendron. Right,* detailed artist of the leaf cushions on the column. Photos by author.

lowing what is called the Permian-Triassic extinction event, the largest extinction event ever in earth's history.

In the Permian, *Pleuromeia sternbergii* was a 2 m (6.5 ft) tall lycophyte with an elongated, unbranched, nonwoody stem that bore helically arranged leaves. The leaves were denser near the base of the

Fig. 4.8. A life reconstruction of the Permian-Triassic *Pleuromeia sternbergii*, which attained a maximum height of ca. 2 m (6.5 ft) and a basal diameter of 22 cm (nearly 9 in). Note the looser, sparser arrangement of leaves and the cone at the top of the trunk, as well as the denser set of leaves and the corm root system at the base of the plant. Reconstruction by Ivo Duijnstee and Hannah Bonner, Creative Commons Attribution-ShareAlike 4.0 International license.

trunk but sparser and held in a looser spiral near the top (fig. 4.8). At the very top of the trunk was a strobilus, or cone, bearing spores of two different sizes. At the base of the trunk was a corm root system. Both the spores of two different sizes and the cormose (bulb-like) root are distinctive features that define the genus *Isoetes* today. Thus,

Pleuromeia is thought to be an ancient relative of the living quillworts, the Isoetales.

In the Lower Triassic, *Pleuromeia sternbergii* began to replace, then overwhelm global plant communities. The remains of this lycophyte are commonly found today in what is called the Zechstein Basin, which extends from eastern England in the west to Denmark in the north, across the Netherlands and Germany, to Poland and Lithuanian in the east. The reason for the takeover of *Pleuromeia* in the fossil plant record was a series of massive volcanic eruptions that produced the widespread flood basalts called the Siberian Traps. The accompanying release of carbon dioxide (CO_2) and other sulfur-rich volatiles formed dust clouds and acid aerosols. These aerosols would have blocked out sunlight and disrupted photosynthesis both on land and in the photic zone of the world's oceans. It is hypothesized that food chains collapsed in the aftermath. A series of other environmental changes occurred, such as an increase in CO_2 levels in the oceans and the atmosphere, a decrease in oxygen in marine water, and global warming. It has been estimated that 90–95 percent of marine species and several groups of terrestrial vertebrates perished in the extinction event that has been called the "Great Dying." In general, though, land plants were less affected than the marine and terrestrial faunas.

Not only was *Pleuromeia sternbergii* able to tolerate these climatic changes but it appears to have even thrived on environmental stress. This lycophyte moved opportunistically into habitats vacated by the dying off of other plant groups around the globe at the end-Permian and into the Triassic, where *Pleuromeia sternbergii* formed coastal communities of low diversity. Later, the spread of woody trees in the global flora began with the resurgence of conifers in the Middle Triassic, followed by cycadophytes and seed ferns. The worldwide return of gymnosperm forests was uneven, taking 10 to 15 million years to bounce back, depending on the region. The lower Mesozoic flora in what is today South China, for example, needed 15 million years to recover completely.

Figure 4.9. A small, sprawling lycophyte called *Lycopodites faleatus*, from the Upper Jurassic Yorkshire flora at Cloughton Wyke, England, photographed here at the Sedgwick Museum in Cambridge. Photo by Smith609 (Verisimilus), Creative Commons Attribution-ShareAlike 2.5 license.

After earth's ecosystems stabilized following the Lower Triassic, lycophytes continued to shrink evolutionarily to a shadow of their former selves in Paleozoic times, diminishing in size, abundance, and habitat dominance (fig. 4.9). Yet, they have managed to survive until today, maintaining a moderate species diversity. Nearly all lycophytes have remained true to their simple morphology and preference for moist and wet habitats for 400 million years.

FURTHER READING

Beals, C. 2017. Flame thrower—Made with moss spores. BealsScience. https://www.bealsscience.com/post/2017/09/12/flame-thrower-made-with-moss-spores.

Benton, M. J. 2005. *When Life Nearly Died: The Greatest Mass Extinction of All Time.* London: Thames & Hudson.

Looy, C. V., J. H. A. van Konijnenburg-van Cittert, and I. A. P. Duijnstee. 2021. Proliferation of isoëtalean lycophytes during the Permo-Triassic biotic crises: A proxy for the state of the terrestrial biosphere. *Frontiers in Earth Science* 9:615370.

Pteridophyte Phylogeny Group I (PPG I). 2016. A community-derived classification for extant lycophytes and ferns. *Journal of Systematics and Evolution* 54:563–603.

Equisetum hyemale

CHAPTER 5

The Horsetail, or Scouring Rush, *Equisetum*

In their quest to settle the West, American pioneers traveled light and had to be resourceful in recruiting nature's bounty for their needs. One useful little plant that they encountered along their way was the scouring rush, which was so tough that it could be used to scrub out pots and pans. The westward settlers had to look no farther than the next body of water for this natural scrubbing pad, for the scouring rush, known botanically as *Equisetum hyemale*, commonly grows on the shorelines of ponds and streams. Other species of *Equisetum* can also be found in moist soils in forests or in more open, seemingly drier areas, such as in meadows or along railroad tracks or roadways (fig. 5.1).

The scouring rush must have been employed for cleaning cookware in Europe as well. It is known as *Zinnkraut*, or "tin herb," in Germany and was useful for polishing pewter as well as for sanding wood. In the nineteenth century, there was a large-scale commercial importation of the rough-stemmed plant from the upper Rhine Valley of Germany through the Netherlands to England, for polishing cabinets and

Figure 5.1. The field horsetail, *Equisetum arvense. Left,* a dense colony of horsetails with wildflowers at the edge of a dry meadow (geological hammer included for scale). *Right,* close-up of one shoot in this colony, showing its characteristic jointed stem and whorls of long, thin branches at the joints. Photos by author.

other wooden objects. Its usage was so common that the trade name of Dutch rush was bestowed upon it. The fine, gritty nature of the scouring rush is also well known in other parts of the world. In Japan, for example, the boiled and dried stems of *Equisetum hyemale* are traditionally used for burnishing wooden handicrafts to impart a smoothness akin to that produced by very fine sandpaper.

The toughness of this thin, tubular plant does not come from having any wood but from having clusters of silica nodules similar to sand grains in its cell walls, especially in the guard cells around the stomata on the upright shoots. One species even has rows of tiny spines made of silica protruding from its stem (fig. 5.2, left). If you rub two stems of any species of scouring rush together, you will hear the scratchy sound made by the silica grains grinding against one another.

Carl Linnaeus was the first to give the scouring rush an official Latin name. In 1753 he put together the words *equus*, for "horse," and *saeta*, meaning "bristle," referring to the thick, unruly shock of thin branches resembling a horse's tail in one species of the genus. As one would expect, this particular species, *Equisetum arvense* (fig. 5.2, right), is so similar to a horse's tail that its common name is mare's tail. In addition to the moniker scouring rush, *Equisetum* is commonly known as the horsetail, or sometimes horsetail rush, to distinguish it from the tail of a true equine. The genus *Equisetum* is thus the name giver for the family, Equisetaceae, and for the order Equisetales.

Figure 5.2. *Left*, tiny spines of silica on the longitudinal ridges along the stem of *Equisetum sylvaticum.* Photo by W. P. de Winter, Creative Commons Attribution-ShareAlike 4.0 International license. *Right*, a colony of *Equisetum arvense* in Newcastle, Northumberland, showing its bushy, horsetail-like form. Photo by MPF, Creative Commons Attribution-ShareAlike 3.0 Unported license.

The living plant *Equisetum* is the only surviving member of the phylum Sphenophyta. Today, there are roughly 15 species in the genus *Equisetum*, most of which thrive in the temperate zone of the Northern Hemisphere. Curiously, they are not native to Australia and New Zealand, although their fossil relatives flourished there in the past. As in many cultivated green spaces all over the world, any horsetails found in Australia and Zealand nowadays will be the bane of gardeners, as they are an introduced, quickly spreading weed that is hard to eradicate. In the paleontological past, the Sphenophyta comprised a hugely diverse and abundant group for millions of years, but their numbers gradually dwindled over time. For this reason, *Equisetum* is truly a sole survivor and living fossil.

With the heady status of a living fossil, you would think that horsetails would be mighty, robust plants, but today's horsetails are slender, mostly low-growing, and gracile. The species *Equisetum arvense* comprises short, narrow, and jointed shoots that arise from underground stems called rhizomes. Most upright shoots are green and do the job that leaves do in most other plants in carrying out photosynthesis. However, in some species there is an additional kind of shoot that bears a cone at its tip and is pale in color (fig. 5.3). In all living *Equisetum* species the leaves are hardly recognizable, for they are reduced to a leaf sheath with black tips wrapped around the shoot and attached at its joint lines. In species that branch, such as *E. arvense*, thin, solid, green branches attached at the stem joints, botanically known as nodes, form the mare's tail.

The horsetail, or scouring rush, is actually not a rush—which is a flowering plant of the family Juncaceae—but a much older and more primitive plant than true solid-stemmed rushes. Instead of having flowers as true rushes do, the scouring rushes reproduce sexually by spores, which are carried in a strobilus, or cone-like structure (fig. 5.4). The ancestors of all horsetails, the cladoxylopsids, first showed up in the Upper Devonian, an evolutionarily intense time when all groups of plants were trying to grow higher, attain arborescence, and form forests (see chapter 3). Nowadays, the horsetail is just an unassuming hol-

Figure 5.3. Two kinds of shoots of *Equisetum arvense* come up in the early spring: the pale-colored shoots with cones and the green, photosynthetic shoots. Both kinds of shoots are divided into segments, with whorls of leaves or branches attached between the joints. Photo by Ted Delevoryas.

low cylinder no more than waist high. However, its close relatives that flourished in the Upper Devonian and Carboniferous, 370–299 million years ago, at heights of 15–20 m (49–65 ft) would have towered over their present-day cousins.

During the Carboniferous, the sphenophytes were able to achieve true tree status. The best-known Carboniferous sphenophyte is *Calamites* (fig. 5.5), which reached a height of 20 m (65 ft) and a diameter of 60 cm (2 ft). The calamites evolved wood cells with thickened cell walls to build tree trunks. Thus they were able to reach the size of what today

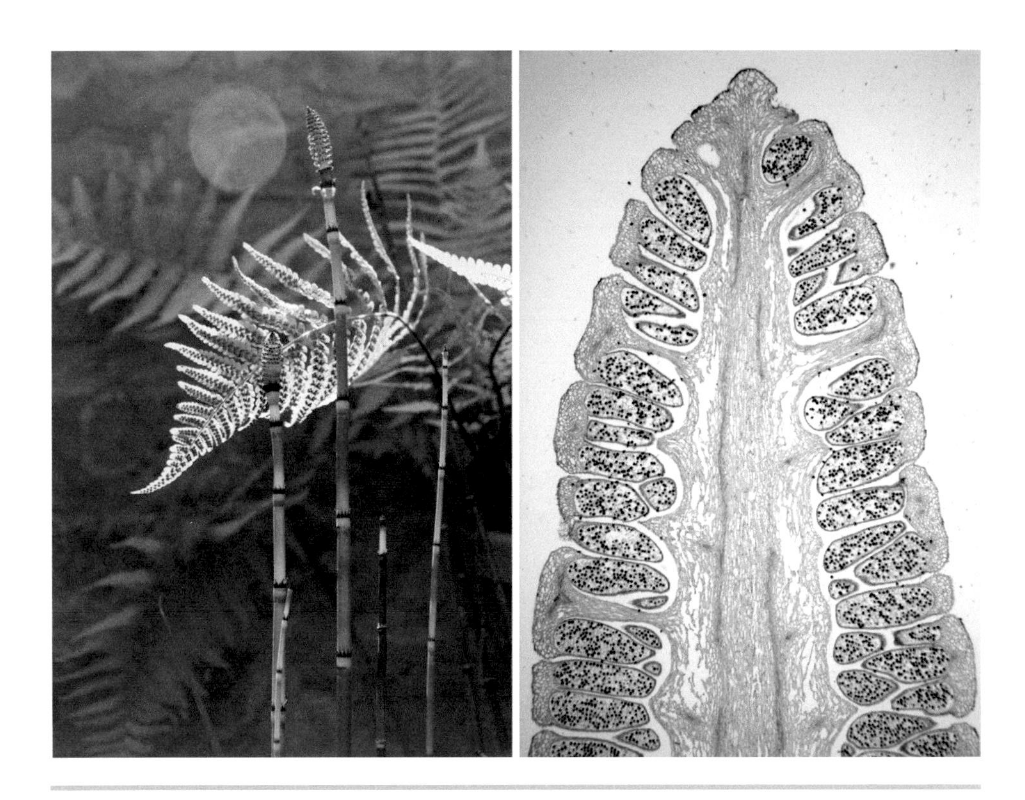

Figure 5.4. *Left,* two kinds of sporing plants: horsetails and ferns. *Equisetum* carries its spores in a strobilus at the top of its stem, while ferns bear their spores on the back of their fronds, appearing here as brown spots. Photo by author. *Right,* A vertical section through a strobilus of *Equisetum arvense,* showing the T-shaped sporangiophores attached to the central axis and the sporangia full of spores, which are stained dark red here. Photo courtesy of Cornell University Plant Anatomy Collection.

would be medium-sized trees. Carboniferous *Calamites* made up the mid-height understory of the coal-forming, peat swamp forests beneath the lycophytes (chapter 4) and cordaites, an extinct group of gymnosperms, which together formed a canopy at 45 m (148 ft).

As in all Sphenophyta, the stems and trunks in the Carboniferous sphenophytes were jointed, with whorls of branches or leaves attached

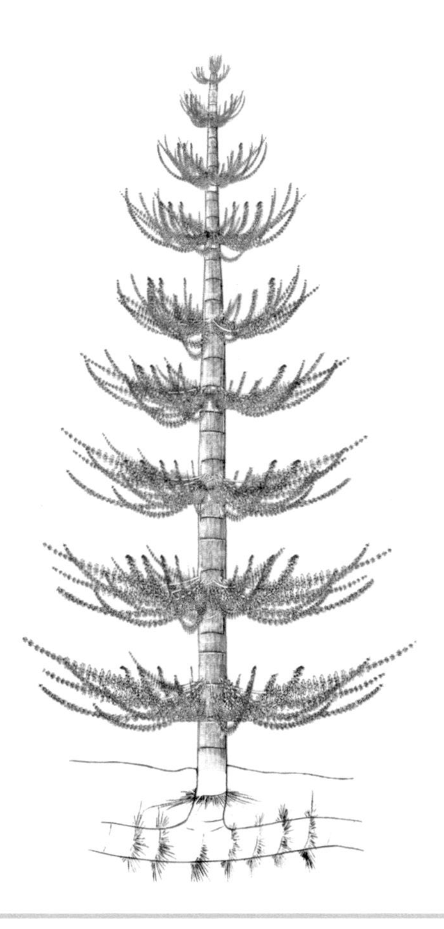

Figure 5.5. A life reconstruction of a *Calamites* tree, which is characterized by its segmented, woody trunk and whorls of branches and leaves. Reconstruction by Falconaumanni, Creative Commons Attribution-ShareAlike 3.0 Unported license.

at the joints (fig. 5.6, left). However, one major difference between Carboniferous *Calamites* and living *Equisetum* is that the leaves of *Calamites* were sometimes broader. In some species, the leaves had a petal shape, lending them the fanciful appearance of daisies (fig. 5.6, center), although the evolution of true flowers would occur 180 million years in the future.

Figure 5.6. Paleozoic and Mesozoic sphenophytes. *Left*, the jointed stem with whorls of narrow leaves of *Asterophyllites longifolius* from the Upper Carboniferous of Eschweiler, near Aachen, Germany. *Center*, whorls of leaves in *Annularia stellata,* having the whimsical appearance of a daisy chain, from the Upper Carboniferous of Saarbrücken, Germany. *Right*, the cast of a robust stem of *Equisetites columnaris* from the Middle Jurassic of the Yorkshire Coast, England. Photos by Georg Oleschinski.

The sphenophytes reached their peak diversity and abundance in the upper Paleozoic. Along with the lycophytes and cordaites, the Carboniferous sphenophytes were an integral part of the coal biomass-forming flora. These ancient calamites may have preferred to settle on the shorelines of lakes and rivers or colonize disturbed areas, much as living horsetails do today. At the end of the Permian, about 252 million years ago, the tall, arborescent members of the plant group faded away, leaving only herbaceous sphenophytes to carry on. These Mesozoic sphenophytes resembled today's horsetails in that they had leaves fused into a sheath wrapped around the stem and lacked wood. They are sometimes given the name *Equisetites* (fig. 5.6, right), referring to their similarity to living *Equisetum*. In fact, the morphological traits found in living *Equisetum* can be traced back 240 million years, to the Middle Triassic. These traits occur all over the plant body, down to the tiny spores, each of which possesses four elaters, thin straps coiled around the spore that spring open to aid in dispersal (fig. 5.7, left).

There are a few differences between extant *Equisetum* and the Mesozoic forms, however. For example, instead of being a gracile, hollow, straw-like stem as in today's horsetail, the Triassic ancestors of *Equise-*

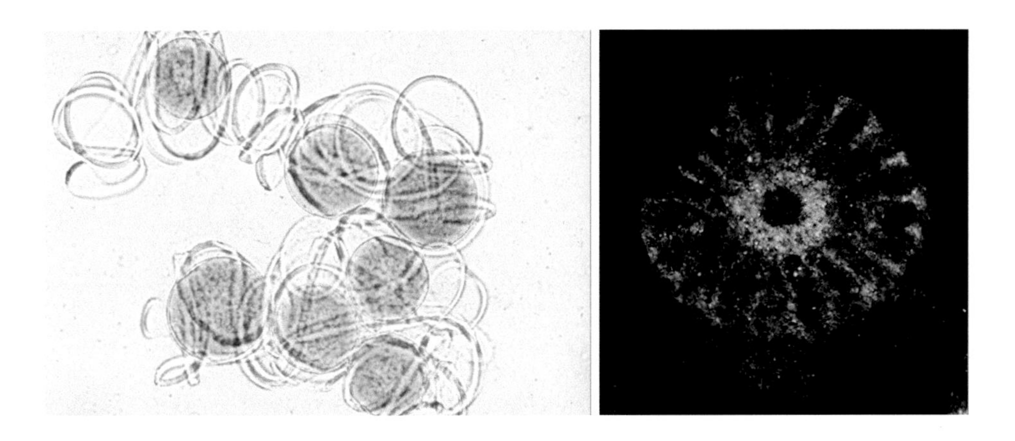

Figure 5.7. Unusual traits in living and fossil horsetails. *Left*, several spores of living *Equisetum*, each with four thin ribbons called elaters coiled around it. In *Equisetum* the spores are green and can photosynthesize. *Right*, the nodal diaphragm of an Upper Jurassic *Equisetum* from Antarctica. Circular disks of tissue at the joints inside the shoots helped to maintain stability in the hollow cylinders in this fossil species. In this case, the nodal diaphragm has taken on the curious form of a wheel, with a central hub, spokes, and rim, and measures ca. 1 cm (0.4 in) across. Photos by author.

tum were sometimes larger, growing to a height of 5 m (16 ft) and a girth of 20 cm (nearly 8 in). In the Jurassic, some species of *Equisetum* began to assume a shorter, narrower form. While upright shoots in living *Equisetum* are hollow tubes, the Mesozoic forms commonly had horizontal partitions of tissue inside the stem at the nodes. These so-called nodal diaphragms were sometimes quite elaborate, resembling a tiny wheel with a central hub, radiating spokes, and a thin, outer rim (fig. 5.7, right).

It was the herbaceous horsetails of the Mesozoic era that led to the lineage that has survived the 190 million years until the present. Today, all horsetails are herbaceous. In the Northern Hemisphere, *Equisetum* species are characteristically low-growing, but some species in Central and South America can reach a height of more than 6 m (20 ft). While the latter are robust and vigorously branched, they do not produce secondary xylem in their stems but are nonwoody.

Actually, it is a wonder that the horsetails did not become extinct in the Mesozoic, for they may have been one of the most favored food plants of the herbivorous dinosaurs. In a world without grass and other angiosperm leaves that make up the natural diet of herbivorous animals today, the plant-eating dinosaurs may have obtained a large portion of their calories, mineral nutrition, protein, and roughage from Mesozoic horsetails. In particular, the fully grown sauropod dinosaurs, huge in size and body mass, needed a steady intake of nutritious, calorie-dense, green food.

Recently, studies have been conducted to analyze the nutritional value of the nearest living relatives of the Mesozoic flora for herbivores. Using the microbe-rich digestive juices of living herbivores, the digestibility or fermentation rate of various plant groups, such as ferns and fern allies, cycads, ginkgoes, and conifers, was precisely measured over three days' time. Normally, this standardized analytical laboratory test, called the Hohenheim Feeds Evaluation Test, is run for only 24 hours to evaluate types of hay and grasses for livestock. However, it was thought that the extra two days of the experiment was a better simulation for the longer gut digestion time of an adult sauropod.

It was expected that the plants considered the quintessential members of the Jurassic flora—the cycads and ginkgoes—would score highest on the test. However, it turned out that the cycads were the worst producers of energy for herbivores and *Ginkgo* was moderate in its energy release. Indeed, the best suppliers of energy by far were the horsetails (fig. 5.8). The digestion of horsetails even exceeded that of grasses, which is the superior food of choice for many herbivores today. Because the analysis of living horsetails shows that they are also high in protein and mineral nutrition, in addition to being a source of high energy, Jurassic *Equisetum* may also have been an especially valuable food plant for young herbivorous dinosaurs.

In today's world, it is mostly waterfowl in the Northern Hemisphere that rely heavily on horsetails. During egg formation, brooding,

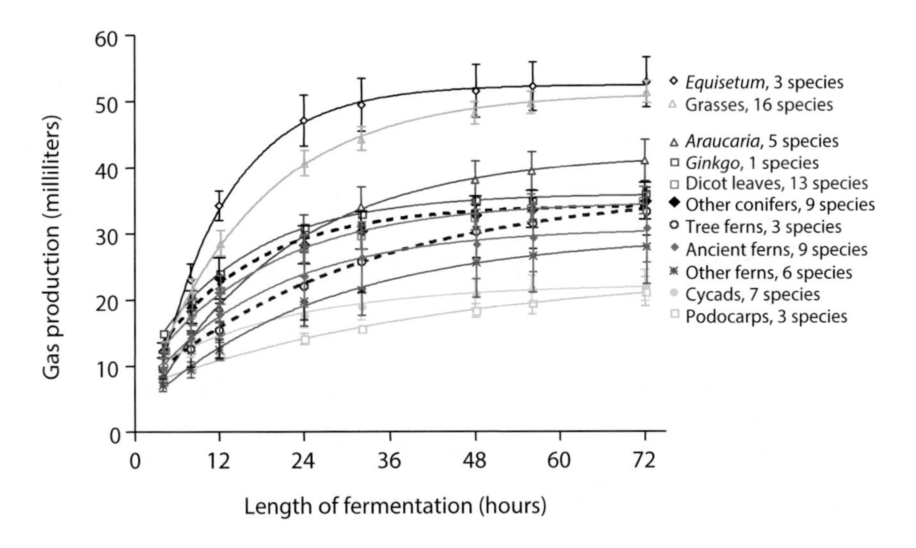

Figure 5.8. Digestibility or fermentation of the nearest living relatives of the Mesozoic flora by herbivores, as measured in laboratory experiments using the Hohenheim Feeds Evaluation Test. Note that horsetails (three species of *Equisetum*) yielded more energy than any other plant group, even grasses. Modified from Gee 2011, also based on data from Hummel et al. 2008.

and reacquisition of body protein after reproduction, birds, including Canada geese, lesser snow geese, pink-footed geese, barnacle geese, and trumpeter swans, consume horsetails growing around boreal lakes in North America. Young geese and cygnets, such as Icelandic pink-footed goslings, have been found to feed exclusively on *Equisetum* for weeks after hatching, utilizing upright stems and rhizome tips as a rich source of amino acids, phosphorous, calcium, magnesium, and potassium.

Larger herbivores, namely, moose, caribou, musk ox, dall sheep, and buffalo, include horsetails as part of a varied plant diet when they feed in wetlands in the summertime. Sheep, cattle, and horses are also known to eat horsetails, but it seems that horses are susceptible to vitamin B1 deficiency due to the thiaminase in *Equisetum*, which inactivates thiamine. On the other hand, cattle in tropical regions have been observed to graze on *Equisetum giganteum* with relish.

People also eat horsetails. In the Japanese and Native American cultures, the early-appearing, pale-colored shoots bearing cones of *Equisetum arvense* are cooked and eaten like asparagus. In Japan, they can also be prepared by simmering them in soy sauce and mirin. In the spring, the people in Siberia enthusiastically look forward to the appearance of young shoots of *E. arvense*, for it is the first edible plant to emerge after the long winter. The high quantity of silica in the *Equisetum* plant body, which is thought to grind down the enamel in teeth, does not seem to pose a serious problem if horsetails are part of a broader mixed diet or if the animal is toothless, as in the case of birds.

The sphenophytes were a major group of plants with robust biodiversity and great abundance in the paleontological past. They arose relatively soon after land plants conquered the earth, and they reached their heyday in the Carboniferous. In the Mesozoic, they may have played an integral role in dinosaur ecosystems as a preferred food source for herbivores, whether small or large, young or old. As hundreds of millions of years went by, however, sphenophyte diversity decreased until only one genus was left—*Equisetum*. Although horsetails are widespread and abundant and grow vigorously today, with more than a dozen species and numerous hybrid forms, the genus *Equisetum* also possesses a long fossil record. These plants display an uncanny close resemblance to ancient species, primitive morphological traits, and a low taxonomic diversity today compared with its relatives in the Mesozoic. For all these reasons, the tenacious survivor *Equisetum* satisfies several criteria for a living fossil.

FURTHER READING

Cornell University Plant Anatomy Collections. 2013. cupac.bh.cornell.edu.

De Winter, W. 2015.The Dutch rush: History and myth of the *Equisetum* trade. *Fern Gazette* 20:23–45.

Gee, C. T. 2011. Dietary options for the sauropod dinosaurs from an integrated botanical and paleobotanical perspective. In *Biology of the Sauropod Dinosaurs: Understanding the Life of Giants*, ed. N. Klein, K. Remes, C. T. Gee, and P. M. Sander, 34–56. Bloomington: Indiana University Press.

Howell, M. M., C. T. Gee, C. Böttger, and K.-H. Südekum. 2023. Digestibility of dinosaur food plants revisited and expanded: Previous data, new taxa, microbe donors, foliage maturity, and seasonality. *PLoS ONE* 18(12): e0291058

Hummel, J., C. T. Gee, K.-H. Südekum, P. M. Sander, G. Nogge, and M. Clauss. 2008. *In vitro* digestibility of fern and gymnosperm foliage: Implications for sauropod feeding ecology and diet selection. *Proceedings of the Royal Society, Series B* 275:1015–21.

Husby, C. 2013. Biology and functional ecology of *Equisetum* with emphasis on the giant horsetails. *Botanical Review* 29:147–77.

Dicksonia antarctica

CHAPTER 6

Ferns and Tree Ferns

The Victorians loved ferns. In nineteenth-century England, artistic renditions of fern fronds adorned all sorts of decorative art in ceramics, glass, metalwork, textiles, woodwork, and sculpture. Ferns were also popular motifs for christening presents and gravestones, thus accompanying the fern-loving populace from cradle to grave. In fact, the Victorians' love of all things pteridological took the form of a societal obsession that was coined *pteridomania*, or fern fever.

The Victorian fever for ferns began in the 1830s but reached its pinnacle in the 40-year period between 1850 and 1890. One reason for the popularity of ferns as a decorative element was that the flat-lying fronds were easily applied and adapted to two-dimensional surfaces. Collecting ferns became a fashionable hobby that spanned the social classes and spawned popular magazines such as *The Phytologist*, in which new discoveries were announced, or books like *The Ferns of Great Britain and Ireland* for the identification of ferns. In fact, the collecting of ferns from the wild became so popular that some rarer species were decimated to the point of near extinction (fig. 6.1). Ferns

Fig. 6.1. *They carried loads of ferns*, drawn by K. M. Skeaping for *The Holiday Prize: A Modern Fairy Tale*, by Elinor Davenport Adams, published in London in 1896.

were not the only objects of interest; also cherished were fern allies such as clubmosses and spikemosses (chapter 4).

Once live ferns were dug up from wild populations or acquired from nurseries, the Victorians commonly transplanted them into what was known as a Wardian case. Developed by Dr. Nathaniel Bagshaw Ward, an English physician and botanist, this growth chamber was an airtight glass house used for growing delicate plants indoors. It became the forerunner of today's terrarium, vivarium, and aquarium. In addition to keeping the ferns' environment moist and humid, the Wardian case also kept out the noxious pollution in the air during Victorian times, which could have been detrimental to the well-being of the ferns.

Nearly a century later, ferns remain quite popular today. In gardens, they are a great horticultural solution for dark, uninviting corners or for heavy, boggy soils. Because many species of ferns prefer soft,

Figure 6.2. Ground-dwelling ferns in the understory of an evergreen conifer forest in the Pacific Northwest. Photo by author.

indirect sunlight, they are also valued as houseplants. In the great outdoors, pteridophyte colonies encountered in fern dells along moist streams or in shady forests with tall trees never fail to dispel the gloom and uplift the spirits of those who come upon them (fig. 6.2). Their delicate and playful forms seem to brighten and soften landscapes. Ferns can even be recruited to serve a mundane purpose such as marking property lines by planting them or encouraging them to grow in hedgerows (fig. 6.3).

Ferns' growth forms are quite varied. The growth habit that most commonly comes to mind is that of a ground-dwelling plant firmly rooted to the soil, creating a kind of green tuft with widespread fronds and reaching no more than waist height. However, ferns can also be sprawling plants with long fronds that creep down slopes or cliffs, such

Figure 6.3. The Old World royal fern, *Osmunda regalis*, forming a hedgerow with pink-flowering heather in the Irish countryside. Photo by author.

Figure 6.4. The Old World forked fern, *Dicranopteris linearis*, or *uluhe* in Hawaiian. *Left*, a sprawling colony of forked ferns scrambling down a volcanically disturbed slope on the Big Island of Hawaii. *Right*, the pinnae of *uluhe* fork in a regular fashion, showing the origin of its common name. Photos by author.

as the Old World forked fern, *Dicranopteris linearis*, which scrambles along volcanically disturbed hillsides in Hawaii (fig. 6.4).

In tropical climates, ferns aspire to grow higher in the competition for light with all the many tall trees in the rain forest community. Thus, some ferns adopt an epiphytic habit, growing high on the trunks of trees to gain more sunlight (fig. 6.5, left). Alternatively, certain ferns can form a tree-like trunk of adventitious roots and leaf bases to hoist their green fronds higher. These arborescent pteridophytes are called tree ferns (fig. 6.5, right).

The most unusual habitat adaptation of ferns may be their embrace of the freshwater realm. Two families—the Salviniaceae and the Mar-

Figure 6.5. Strategies of ferns to gain more sunlight by moving their leaves upwards in the canopy. *Left*, the epiphytic ferns of the oakleaf basket fern, *Aglaomorpha quercifolia* (formerly *Drynaria quercifolia*), nestled in the branches of a bull kauri, *Agathis microstachya*, in North Queensland, Australia. The basket ferns trap organic debris at the base of their fronds to supply themselves with mineral nutrients that would otherwise come from the soil. The two evergreen trees are the so-called twin kauris on Lake Barrine, estimated to be 50 m (164 ft) high and more than 1,000 years old. *Right*, looking up at the fronds of a tree fern of *Cyathea* in the rain forest of North Queensland, Australia. Photos by author.

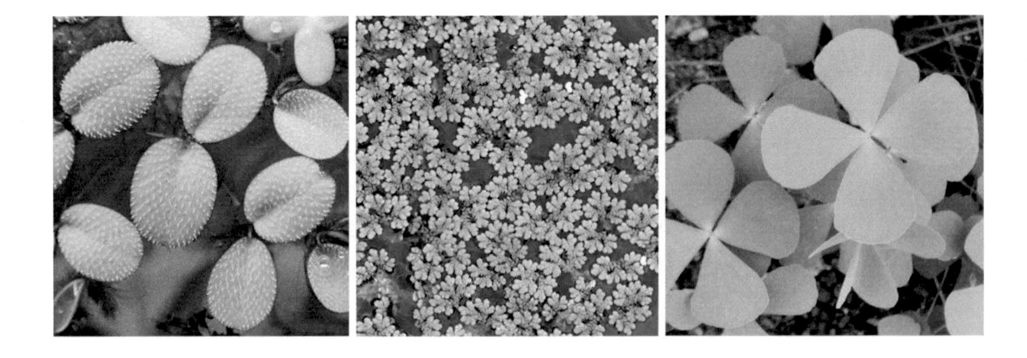

Figure 6.6. Water ferns. *Left*, the common mossfern, *Salvinia minima*, here at the Botanical Garden of the University of Wrocław in Poland. Photo by Krzysztof Ziarnek, Creative Commons Attribution-ShareAlike 4.0 International license. *Center*, the red azolla, *Azolla rubra*, native to Australia and New Zealand. Photo by Jeremy Rolfe, Creative Commons Attribution-ShareAlike 4.0 International license. *Right*, the European water clover, *Marsilea quadrifolia*. Photo by Vinayaraj, Creative Commons Attribution-ShareAlike 3.0 Unported license.

sileaceae, comprising roughly 77 species—make up the water ferns (fig. 6.6). The genera *Salvinia*, generally known as the watermoss, and *Azolla*, commonly called the duckweed fern, are both small ferns that float free on the surface of quiet ponds or bayous. *Salvinia* and *Azolla* water ferns are made up of chains of buoyant leaves connected by sections of thin stem. *Marsilea* ferns, on the other hand, tend to root along the shallow shoreline of ponds and send up foliage that looks like four-leaf clovers. In aquaria, *Marsilea* ferns will grow completely submerged, forming a low green carpet that offers a nice microhabitat for freshwater shrimp or other small crustaceans.

Despite the variety of their growth forms, all ferns have young leaves that uncoil or unfurl, looking very similar to the scroll at the head of a violin, for which reason young fern fronds are often called fiddleheads (fig. 6.7). Tightly coiled, freshly harvested fiddleheads of some species can be cooked—boiled, steamed, or sautéed—then tossed with greens in a salad or served on a bed of pasta or risotto.

Figure 6.7. A young, unfurling fiddlehead of the king fern, *Angiopteris*, growing in the Taichung Botanical Garden in Taiwan. Photo by author.

The arborescent habit, lacy fronds, and unfurling fiddleheads of tree ferns sometimes evoke a prehistoric feeling. Indeed, while more than 10,560 living species of ferns are known today, ferns are among the most ancient terrestrial plants on earth. The geological period following the Devonian—the Carboniferous—is sometimes called the

Figure 6.8. The Asian royal fern, *Osmunda japonica*, in a subtropical forest in southwestern China. Photo courtesy of Hans Hagdorn.

Age of Ferns. While the swamp forests 359–299 million years ago were filled with a plethora of different kinds of plants, including giant trees of lycophytes (chapter 4), the now extinct cordaites and seed ferns, and sphenophytes (chapter 5), the true ferns underwent rapid radiation in the Carboniferous. However, many of the fern fronds found in this period bear a passing, superficial resemblance to the fern foliage of the present day.

One of the most remarkable fern lineages in deep time with an extensive fossil record is the order Osmundales. This group of ferns first appeared in the Permian and rapidly diversified to continue strong into the Mesozoic. *Osmunda* continues to survive today. An estimated more than 100 species of Osmundales in the paleontological past have been described from just fossil stems and rhizomes. In contrast, 25 species in the family Osmundaceae are known in the present day. One of the most widespread members of the living Osmundaceae is *Osmundastrum cinnamomeum*, the cinnamon fern, which is native to eastern North and South America, Europe, Asia, and southern Africa. Also well known are *Osmunda regalis* (see fig. 6.3), the Old World royal fern, found in Europe, Africa, and Asia, and *Osmunda japonica* (fig. 6.8), the Asian royal fern, native to East Asia.

The order Osmundales possesses a number of typical characters that aid in its recognition in the fossil record. One is its stem anatomy, which includes distinctive C-shaped leaf traces. Another is frond size, shape, and structure. The fossil leaves of osmundalean ferns are relatively easy to recognize because their robust, bipinnate fronds are quite similar to those of living foliage (fig. 6.9).

In the paleontological record of the Permian, osmundalean ferns appear mainly as anatomically preserved stems. By the Upper Triassic, however, the evolutionary development of the genus *Osmunda* was well under way. Fern fronds found in 220-million-year-old rocks in Antarctica are virtually identical in form and size to the sterile fronds of the living species of *Osmunda claytoniana*, the interrupted fern. The unusual common name of *O. claytoniana* is attributed to the "interruption" of the normal-sized, green-colored pinnae on a frond by small, short, brown-colored fertile pinnae with clusters of the spore-bearing sporangia. This distinctive "interruption" of the sterile pinnae by reproductive material appears exactly the same in the fertile frond of the fossil fern. Thus, the *Osmunda claytoniana*–like fern described

Figure 6.9. Fronds of osmundaceous ferns, fossil and extant. *Left*, a 150-million-year-old frond from the Upper Jurassic flora at Hope Bay, at the northernmost tip of the Antarctic Peninsula. *Right*, a frond of the living fern *Todea barbara*. Photos by author.

from the Upper Triassic is considered to be the oldest known fossil record of any living fern species.

Another spectacular example of fossil Osmundaceae is found in the Lower Jurassic of southern Sweden. There, a fern stem was mineralized while still alive by calcium-rich hydrothermal brines. The fossilization of the fern rhizome is so exquisite that cytoplasm and nuclei inside the individual cells are well preserved. The mineralization process was so amazingly fast that chromosomes inside the nuclei were caught in the various stages of cell division. The presence of a single root per leaf trace in the fossil fern points to an affinity with the living genus *Osmundastrum*, which lives on today as the cinnamon fern. From the size of the genetic material in the nuclei, it can be confirmed that the fossil fern and the living cinnamon fern share the same number of chromosomes. This suggests that neither an increase nor a decrease in chromosomes has occurred in this species for the last 180 million years. Thus, the longevity and genomic stasis of Osmundaceae ferns are evident in the 220-million-year-old fern from Antarctica, which is nearly identical to the living interrupted fern, *Osmunda claytoniana*. Similarly, a genetic stasis on the species level is echoed by the 180-million-year-old fern from Sweden, which is virtually similar to the living cinnamon fern, *Osmundastrum cinnamomeum*, down to the micrometer-scale size of its cell nuclei.

In the end, ferns are living fossils. This is not because they represent the few last surviving remnants of an ancient lineage but instead because they are one of the oldest major plant groups. Their first appearance in the history of life was 390 million years ago, a "mere" 40 million years after plants started conquering land. Ferns were at the vanguard in experimenting with the arborescent habit, building tree ferns in an attempt to grow taller than their neighbors in the competition for sunlight. Lastly, fern species can have extreme longevity, extending over hundreds of millions of years, and remain constant in their morphology of all plant organs, from the internal structure of their stems to the form of their fertile leaves to the size of their nuclei and the number of their chromosomes. In short, the Victorians were justified in their fever for ferns, not only as beautiful decorative motifs but as amazing plants from extraordinarily deep time.

FURTHER READING

Bomfleur, B., S. McLoughlin, and V. Vajda. 2014. Fossilized nuclei and chromosomes reveal 180 million years of genomic stasis in royal ferns. *Science* 343:1376–77.

Boyd, P. D. A. 2002. Pteridomania—The Victorian passion for ferns. http://www.peterboyd.com/pteridomania.htm.

Moran, R. C. 2009. *A Natural History of Ferns.* Portland: Timber Press.

Phipps, C. J., T. N. Taylor, E. L. Taylor, N. R. Cúneo, L. D. Boucher, and X. Yao. 1998. *Osmunda* (Osmundaceae) from the Triassic of Antarctica: An example of evolutionary stasis. *American Journal of Botany* 85:888–95.

PART III

CHAPTER 7

Bearing Seeds and Woody Cones

Sex is always a bit touch and go, even between botanical partners. Plants like lycophytes, horsetails, and ferns are known as cryptogams, which means that their means of reproduction is hidden. Their reproductive parts are tiny, consisting of separate sperm-bearing and egg-bearing sex organs on small gametophytes. For the sperm to reach the eggs, they need to swim through water. Thus, the first land plants, as well as their relatives that live on today, required water to have sex and carry out genetic recombination.

The union of sperm and egg in a cryptogam results in a small plant, the sporophyte, that grows directly on the gametophyte. The sporophyte is also dependent on water because of the thinness of its outer tissue and its need to stay moist. Thus, the sporophyte must avoid getting dried out. This is why many lycophytes, horsetails, and ferns prefer moist habitats. Imagine, then, the freedom from dependence on water that is offered by morphological innovations preventing desiccation, such as the evolution of the seed (fig. 7.1).

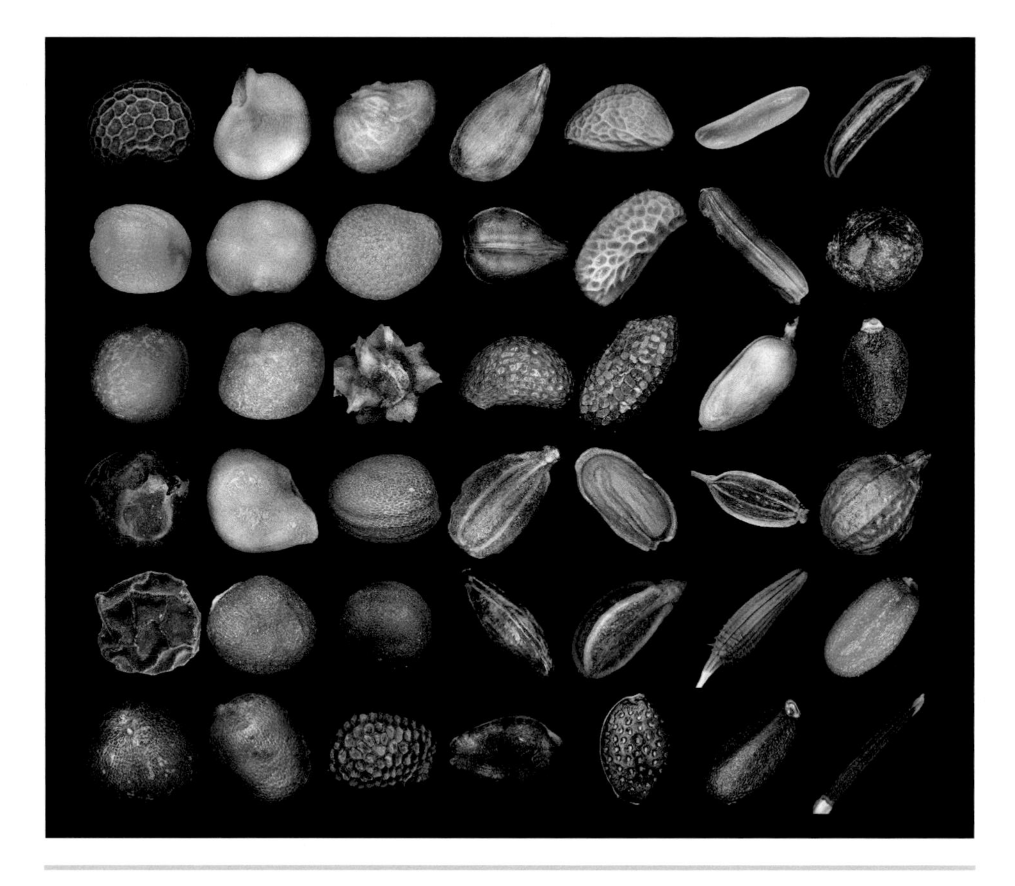

Figure 7.1. Close-ups of seeds from a variety of living plants. *Top row*, poppy, red pepper, strawberry, apple, blackberry, rice, caraway. *Second row*, mustard, eggplant, groundcherry, grape, raspberry, red rice, patchouli. *Third row*, fig, goji, beet, blueberry, golden kiwi, rosehip, basil. *Fourth row*, pink pepper, tomato, radish, carrot, Brompton stock, dill, coriander. *Fifth row*, black pepper, white cabbage, napa cabbage, sea buckthorn, parsley, dandelion, shepherd's purse. *Bottom row*, cauliflower, radish, kiwi, sweet grenadilla, passion fruit, melissa, marigold. Photo by Alexander Klepnev, Creative Commons Attribution-ShareAlike 4.0 International license.

The seed was a truly advantageous innovation. This evolutionary innovation occurred after plants had conquered the transition from the sea to damp habitats on land and helped plants to disperse their offspring into dry environments. Depending on the thickness and the water impermeability of the seed coat, it can keep the living tissue inside dry and intact for long periods of time. An extremely tough outer coat is, for example, the reason why seeds of the sacred lotus will stay alive and even germinate 1,300 years later (see chapter 16). Within the seed coat of many seeds is an energy-filled, starchy tissue called endosperm, which nourishes the plant embryo until conditions are right for germination. After all, what is a seed but a baby plant in a box with its lunch?

Ancient Seeds

The first seed-like structures showed up in the Upper Devonian, about 370 million years ago, at the time when various plants were toying with different strategies to attain greater height (see chapter 3). The name of the oldest seed-bearing plant in earth history, *Elksinsia polymorpha*, honors the place where it was found, Elkins, West Virginia. The geologically slightly younger early seed *Archaeosperma* was named to reflect its place in paleobotany, *archaeo* being derived from the Greek word *arkhaios*, meaning "ancient," and *sperma* stemming from the Latin *sperma*, for "sperm" or "seed." In the early seed plants, the seeds were borne directly on the stem. They developed in ones or in twos and were surrounded by a loose, cup-shaped structure known as a cupule. The cupule commonly had a deeply lobed margin. The seed was enwrapped by a second tissue that later evolved into an integument, the forerunner of the seed coat.

The reproductive and ecological advantages of seeds must have caught on. From this point on, plant evolution continued to refine the morphological, anatomical, and physiological functions of seeds. Except for one end, which was left open, the seed was protected by an enveloping integument. The opening allowed pollen to enter so that the sperm cells could fertilize the egg cell within. Later, in cycads and conifers, a small pollen chamber developed just inside the opening that exudes a droplet of sticky fluid to capture pollen grains and draws them to the egg as the fluid dries and retracts into the pollen chamber. In this way, the plant produces its own fluid to facilitate the union of

Figure 7.2. Large seeds dangling from the fern-like fronds of a seed fern, reconstructed in a lifelike diorama of ancient life in the Museum Mensch und Natur in Munich, Germany. Photo by Petter Bøckman, Creative Commons Attribution-ShareAlike 4.0 International license.

sperm and egg and thus is no longer dependent on the randomness of rainwater or limited to growing in a moist environment.

With the advent of the seed habit, the plant world became divided into free-sporing plants and seed plants. Today, lycopods, horsetails, and ferns (see chapters 4–6) have retained their spore-bearing type of reproduction, whereas cycads, ginkgoes, conifers, and angiosperms (see chapters 8–16) embraced the new morphological innovation and became seed-bearing plants. At the time of this dramatic split, approximately 370 million years ago, the earliest seed plants had characters intermediate between those of free-sporing plants and those of seed plants. Like ferns, they had lacy fronds, but instead of reproducing by spores as ferns do, they bore seeds (fig. 7.2). Paleobotanists thus refer to this large, heterogeneous group of plants as seed ferns. While in the light of plant evolution they are viewed as a transitional plant group with intermediate characters, seed ferns were actually a species-rich and successful group of plants that made up the understory of the world's forests. They existed for more than

300 million years, throughout most of the Paleozoic and for all of the Mesozoic. Many seed fern lineages went extinct at the same time the dinosaurs did, at the end of the Cretaceous, but a few managed to make it into the Cenozoic.

Cones

A seed cone is similar to a basket of eggs, a practical and protective wooden structure for carrying a plant's fragile future offspring. Alternatively, in view of the many horizontal levels it has, perhaps it would be better to compare a cone to a multitiered cake stand with highly prized petit fours on each level. However, the scales of a cone commonly swirl around a central axis in a spiral manner and bear small reproductive units such as spores, seeds, or pollen.

As indicated in the term *cone*, some plant cones, especially pine cones, are generally shaped like a geometric cone, with a narrower top and a wider bottom. However, a cone is a generalized structure; cones are found throughout the plant kingdom. The horsetail, *Equisetum*, has cones that bear spores (see chapter 5), while cycads and conifers have woody pollen and seed cones. Even some flowering trees, such as alders, have seed cones. It is likely that cones developed independently in each of these groups, because a cone is one of the most compact structures that can be achieved by arranging elements around a central axis.

Cycad Cones

In cycads, cones are produced at the top of the trunk, borne in the center of a crown of green leaves (fig. 7.3). They are generally massive, reaching lengths in excess of 76 cm (30 in). Individual cycads produce either pollen (male) or seed (female) cones. In other words, cycads segregate the sexes to separate plants to facilitate cross-fertilization and to prevent self-pollination. Inside the female cones are the ovules, which

Figure 7.3. Seed cones in the crown of a cycad at the Huntington Botanical Gardens, San Marino, California. Photo by author.

are unfertilized seeds. Cycad seeds often have a hard seed coat that is commonly bright orange in color (see fig. 8.7, right). After the ovules are fertilized by pollen, they will develop into viable seeds. However, there is one cycad, *Cycas*, in which the origin of the seed-bearing cone scales from leaves is still evident. In the species *Cycas revoluta*, the seed-bearing cone structure, which is technically called a megasporophyll, retains a leafy fringe, which serves as a reminder of its evolutionary origin from a cycad frond (see fig. 8.8).

Conifer Cones

Among the conifers, the best-known cones are probably those of pines. In members of the pine family, individual trees have both female and male cones (fig. 7.4), unlike the cycads, which have only one or the other sex. The pollen cones of conifers are small and yellow. They stay on the tree for just a short time, then fall off once their pollen has been shed. Seed cones, on the other hand, are borne by the tree for another 1.5 to 3 years after the ovules have been fertilized. The seed cones of pine are compact woody structures with overlapping cone scales arranged in a spiral. Their tapered, cone shape makes them the quintessential conifer cone for most people.

The woody scales of a pine seed cone open and close depending on how wet the cone is. When the cone is wet, the scales close up, and when it is dry, they open. This ensures that the seeds, which commonly have a thin, papery wing on them, will remain dry and be able to disperse in dry weather by fluttering off in the wind. This special property of pines to open and close their cones has led some people living in temperate climates to hang a pine cone outside to give notice of wet or dry weather.

In pine species that inhabit forest communities in which wildfire plays an important ecological role, their cones have developed adaptations to release their seeds at the right time for optimal germination. In these species, the cone scales are sealed shut with the pine's own resin until the heat of a wildfire melts the resin and opens up the cone. Seeds released after a fire face less competition for seed germination and growth of young seedlings. There is also increased sunlight after a fire, as well as a short-term increase in nutrients in the soil from fire ash. Pine species well known for their serotinous cones—ones that remain closed on the tree until there is an ecological trigger for seed release—are lodgepole pines and jack pines in North America (fig. 7.5).

Figure 7.4. Cones of both sexes on the same pine tree. The yellow pollen cones near the top of the branch are mature and shedding masses of yellow pollen on the young, still-developing seed cone below. The pollen cones in a cluster at the lower left are immature. Photo by author.

Many other groups of conifers also bear woody cones. Pines, for example, have hard, woody seed cones and smaller, softer pollen cones similar to araucarias, kauris, and the Wollemi pine (chapter 10). The seed cones of the Pinaceae and Araucariaceae are similarly constructed,

Figure 7.5. A charred seed cone of lodgepole pine, *Pinus contorta*, one month after the disastrous Maple Fire in Yellowstone National Park in August 2016. This serotinous cone is still attached to the tree, but it has newly opened as a result of heat from the wildfire to disseminate its seeds. Photo by Jim Peaco, retrieved from the Yellowstone Digital Slide Files.

with spirally arranged cone scales that hold the seeds. One difference between the seed cones of these two families is that those of pines are held pointing downward toward the ground (see figs. 7.4 and 7.5). In contrast, the seed cones of the Araucariaceae are held upright on branches (fig. 7.6). More importantly, however, all pinaceous cone scales bear two seeds, while araucariaceous cone scales have only a single seed. This would make it easy to distinguish the botanical affinities of seed cones in the fossil record if one could look inside the cone.

One way to look inside a woody cone would be to break off the cone scales individually. However, this painstaking task would ruin the integrity of the cone. While the number of seeds per cone scale may become apparent, it is doubtful that the exact spiral arrangement of the cone scales and seeds would become visible. In the case of fossil cones that have been mineralized into solid, stone-hard objects, thin sections could be sliced through the cones with a rock saw, but this would permanently destroy the precious fossils. Fortunately, modern technology has provided us with a way to look at the internal structure

Figure 7.6. Young seed cones of the family Araucariaceae at the Huntington Botanical Gardens, San Marino, California. *Left*, *Araucaria cunninghamii* cones. Photo by author. *Right*, Close-up of a seed cone of *Wollemia nobilis*. Photo by Danielle Rudeen.

of recent and fossil cones nondestructively. The use of high-resolution x-ray computed tomography, or microCT, allows for the study of the internal construction of conifer seed cones (fig. 7.7). The nondestructive microCT method allows the conifer cone, whether recent or fossil, to remain intact. Three series of x-ray images, amounting to thousands of individual photos, are produced that can be used to study the internal structure of the cone. In pine cones, for example, the seeds are not arranged in a single continuous spiral, contrary to expectations. Rather, there are three seed spirals, each of which wraps around the cone multiple times. Single-image slices through the cone can provide further evidence of the three sets of seed spirals. Moreover, it can be clearly observed that each cone scale bears two seeds, a distinctive feature of the Pinaceae. Although microCT has already shown itself to be a powerful scientific tool, there is no doubt that it will continue to be important in

Figure 7.7. A seed cone of *Pinus pinea*, known as the umbrella pine in the Mediterranean. *Left*, side view taken with conventional photography. Photo by Georg Oleschinski. *Center*, side view of the three sets of seed spirals that wrap around the cone multiple times, created by microCT and 3D reconstruction using Avizo software. From Gee 2013, Attribution-NonCommercial-ShareAlike license. *Right*, cross section through the seed cone, showing the three sets of seed spirals, marked here with red arrows, as well as the two seeds in each cone scale. Photo by author.

botany and paleobotany by providing deeper insight into the morphology, anatomy, and architecture of cones and other plant parts.

FURTHER READING

awkwardbotany. 2021. The serotinous cones of lodgepole pine. https://awkwardbotany.com/2021/12/08/the-serotinous-cones-of-lodgepole-pine/.

Botanic Wonders. n.d. Cycad cones. https://www.botanicwonders.com/Feature-Cycad-Cones.html.

Gee, C. T. 2013. Applying microCT and 3D visualization to Jurassic silicified conifer seed cones: A virtual advantage over thin-sectioning. *Applications in Plant Sciences* 1:1300039.

Gee, C. T., R. D. Dayvault, R. A. Stockey, and W. D. Tidwell. 2014. Greater paleobiodiversity in conifer seed cones in the Upper Jurassic Morrison Formation of Utah, USA. *Palaeobiodiversity and Palaeoenvironments* 94:363–65.

Cycas revoluta

CHAPTER 8

Cycads

For many, it is a toss-up, really, whether cycads or tree ferns best embody the spirit of a prehistoric plant. On one hand, cycads, with their stumpy stature and palm tree–like fronds, appear side by side with long-necked dinosaurs in many Mesozoic dioramas. On the other, tree ferns, with their long, narrow trunk, topknot of foliage, and gracile, coiled young fronds, seem like something straight out of *The Flintstones*. In any case, today's cycads have managed to retain their exotic primitive look since their appearance on earth as far back as the Upper Carboniferous or Permian, 300 million to 280 million years ago (fig. 8.1).

The Mesozoic is called the Age of Dinosaurs by some, but in the view of plant paleontologists it is more truly called the Age of Cycads. It was during this geological time period that cycads became a major force in the global flora. Not only did cycads increase in numbers but they rapidly evolved and grew in species diversity. Then at the end of the Mesozoic, 66 million years ago, when dinosaurs went extinct, the cycads slipped

Figure 8.1. A grove of *Cycas revoluta* cycads in the Japanese Garden at the Huntington Botanical Gardens, San Marino, California. Photo by author.

past them. The cycads survived the Cretaceous-Paleogene extinction event and even thrive on earth to this very day.

The cycad that most people are acquainted with is *Cycas revoluta*, the sago palm, which has a relatively short and stout woody trunk and a crown of dark green leaves. However, other species of *Cycas*, such as *C. circinalis*, the queen sago, bear their leaves atop a tall, tapering trunk, making them look like palm trees that are a bit stout near their roots (fig. 8.2). The tallest cycad is thought to be *Lepidozamia hopei*, or Hope's cycad, which is native to Queensland, Australia. This giant cycad can reach a height of 20 m (66 ft), and its stem can be as thick as 50 cm (1.6 ft). The two most massive cycads are *Encephalartos laurentianus*, the Kwango giant cycad, at home in northern Angola and southern Congo, and *Macrozamia dyeri*, known locally as *djeeri* in

Figure 8.2. *Cycas circinalis* is the only gymnosperm in the Sri Lankan flora, where it is found naturally. Photo from 2009 by Jee & Rani Nature Photography, Creative Commons Attribution-ShareAlike 4.0 International Public license.

Figure 8.3. Similarities in the general appearance of a cycad, a palm tree, and a tree fern. *Left*, the cycad *Dioon spinulosum* at the Brooklyn Botanical Garden in New York. Photo by author. *Center*, the date palm, *Phoenix dactylifera*, in Balochistan, Pakistan. Photo by Yasir Dora, Creative Commons Attribution-ShareAlike 4.0 International license. *Right*, a tree fern of *Cyathea* in Mount Kinabalu National Park, Malaysia. Photo by Bernard Dupont, Creative Commons Attribution-ShareAlike 2.0 Generic license.

Western Australia, where it grows. Both of these species can achieve a trunk diameter of 1 m (3.3 ft).

Cycads are commonly mistaken for palms, which is hardly surprising since both have feather-like leaves at the top of a woody trunk (fig. 8.3, left and center). However, cycads and palms are quite different from each other. Cycads are gymnosperms with seeds usually borne by a woody cone, while palms reproduce by producing small, numerous, and fleshy fruits such as dates or larger fruits like coconuts. Cycads can also seem similar to tree ferns (fig. 8.3, left and right). Unlike ferns, which can have highly divided fronds, cycad fronds are commonly only once or perhaps twice pinnate. Moreover, cycad leaves are stiff, leathery, and often prickly, while tree fern fronds are soft and delicate, with gracile, coiled young leaves. More importantly, they are also quite different in their mode of reproduction, because tree ferns produce spores.

Cycads grow all over the world, but they are particularly common in the subtropics and tropics. There are only 10 genera, and the greatest number of species are native to South and Central America. Yet they are also found in North America, Australia, Africa, and Madagascar, as well as across Asia. Cycads are equally at home in many habitats, from the understory of wet rain forests to drier environments such as deserts or open eucalyptus woodlands (fig. 8.4). They are easy to spot from a distance owing to their short, cylindrical trunks with stiff, spreading fronds on top.

Cycads are notoriously slow growers. Most cycads grow so slowly that it can take many years for their woody stem to become tall enough to become visible as a stumpy trunk under their radiating crown of fronds. In some species, the stem even grows half-buried in the ground and never really develops a distinct tree-like form.

The slow growth rate of cycads is also reflected in how infrequently they produce leaves. Unlike plants that continuously develop leaves,

Figure 8.4. Two very different habitats with naturally occurring cycads in northeastern Australia. *Left*, a rain forest with a cycad in the understory. *Right*, a dry open grassland with white-barked eucalyptus trees and low-growing cycads. Photos by author.

Figure 8.5. The slow growth of cycads makes them suitable for use in decorative settings. *Left, Cycas revoluta* cycads are often planted in ceramic pots as grand accents to hotel entrances in China, shown here as one of two on either side of a hotel portal in Shiyan, Hubei Province. *Right, Zamia furfuracea,* a species native to Mexico, lends an exotic touch to an ordinary sidewalk in Cairns, northeastern Australia. Photos by author.

cycads usually grow a new flush of several leaves all at once, but it may be a long time before these new leaves show up. In the wild, cycads usually grow a new cycle of leaves once a year, while under cultivation they may produce the next crown of leaves every two years. In Asia, solitary, slow-growing cycads are commonly set into large flowerpots and used as striking ornaments to frame the portals of grand hotels (fig. 8.5, left). This creates a regal look and has the added advantage of avoiding the trouble of frequent transplanting into larger containers. Their leisurely pace of growth makes cycads quite easy to keep as potted bonsai, decorative ornamentals in small gardens, or evergreen shrubs in plant beds on sidewalks (fig. 8.5, right). They can also be incredibly long-lived; the oldest individual cycad is estimated to be about 2,000 years old on the basis of the number of leaf scars on the woody trunk.

The evergreen leaves of low-growing cycads might seem to be a convenient and tasty source of nourishment for plant eaters. However, cycads have evolutionarily adapted to defend themselves. Most cycad fronds are rigid and tough, and they can also be extremely prickly or spiny. While a "normal" cycad, such as *Cycas revoluta*, has leaflets that are pointed only at the tips, there are also extreme cycads, such as *Encephalartos horridus*, the Eastern Cape blue cycad, that have truly horrifically spiny leaves (fig. 8.6). In their native South Africa, Eastern Cape blue cycads are purposely planted to serve as a living fence to keep livestock in and predators out. This species of cycad is not very tall, reaching only 0.9 m (3 ft) in height, yet it must form a visually formidable, physically painful thicket to hungry carnivores outside the prickly pen.

In addition to structural defenses, cycads are also known to employ chemicals as self-defense against herbivores. Parts of all cycads contain toxins, although poisoning most commonly occurs through ingestion of the leaves or seeds. *Cycas revoluta* and *Zamia pumila*, for example,

Figure 8.6. Spiny leaflets of *Encephalartos horridus*, the Eastern Cape blue cycad, photographed here at the Jardin botanique de la Villa Thuret in Antibes, France. Photo by Daderot, available at Wikimedia Commons.

contain the compounds cycasin and macrozamin, which are both carcinogens and neurotoxins. Range livestock that feed unsuspectingly on cycads will develop "zamia staggers," in which their limbs are cumulatively and irreversibly paralyzed. Continued ingestion of the toxic cycad plant parts will end in the animal's death.

Despite their toxicity, cycads are harvested by humans as food. Like some palms, cycads contain a spongy, starchy tissue inside their trunk that is used to produce sago, a major staple food in the lowland areas of New Guinea and the Maluku Islands. Sago must be leached of its poisonous substances, however, before it can be consumed. The starchy tissue from the cycad stem, or sometimes from seeds, is first dried, then ground into a fine powder. After boiling it in water, draining it will leach away the toxic compounds and leave behind the starch. The starchy matter is then repeatedly dried and pounded until a fine powder is produced that has as little toxin as possible. Hence, *Cycas revoluta* is commonly called the sago palm, while the edible properties of the chestnut dioon of eastern Mexico are enshrined in its Latin name, *Dioon edule*.

When ripe, the seeds of many cycads are brightly colored. For example, the seeds of *Macrozamia johnsonii* are deep orange (fig. 8.7). In this species as in many other cycads, the seeds are borne in large, woody cones at the top of the plant. The cone scales that make up the seed cones are each armed with a robust spine, which discourages premature consumption of the seeds. However, when the cones are mature, they break open and the seeds spill out, creating a spectacular tumble of bright orange seeds. Birds are generally attracted by orange and red colors, and in the case of cycad seeds, they are thought to serve as seed dispersers. Cycad seeds are covered by a fleshy tissue that offers a food reward; birds have been observed to pick up a cycad seed, fly away, and consume it elsewhere, digesting only the juicy seed covering. If the seed itself passes unscathed through the bird's digestive system and is expelled intact, it is a win-win situation for both bird and cycad: the bird enjoys a succulent meal, and the cycad's offspring is dispersed safely far from the parent plant.

Figure 8.7. *Macrozamia johnsonii*, native to New South Wales, Australia, growing here at the Huntington Botanical Gardens in San Marino, California. *Left*, a short-trunk example of *Macrozamia johnsonii* with long green fronds. *Right*, three cones, one of which has broken open to expose its brightly colored seeds, nestled between older leaves and a flush of younger fronds. Photos by author.

In present-day Australia, a variety of birds have been observed to eat cycad seeds. They range from smaller birds, such as the native parrots, pigeons, cockatoos, silvereyes, butcherbirds, and ravens, to larger, flightless birds like the emus. Also attracted by this source of food are mammals, including endemic marsupials such as kangaroos, wallabies, and brushtail possums, as well as unfamiliar animals such as quokkas or western quolls, also called chuditch.

Yet some propose that the weight and size of cycad seeds point to evolution in conjunction with the large herbivorous animals of the Ice Age, the so-called Pleistocene megaherbivore fauna. These large animals could have easily dispersed the heavy seeds long distances in the paleontological past. They would have included giant ground birds larger than the present-day emus and *Diprotodon*, a rhino-sized marsupial quadruped, which have since gone extinct. However, today's cycads likely do not suffer that seriously from the loss of megaherbivore dispersers and the lack of long-distance dispersal. The heavy seeds of the Australian cycad *Macrozamia miquelli*, for example, whose seedlings rarely germinate more than 1.5 m (5 ft) from the mother plant, seem to do just fine sprouting up near the parent plants to form large single-species groves or colonies.

About half of all cycads provide seeds, for individual cycad plants are either male or female. While male cycads may also bear cones, their cones produce pollen. In *Macrozamia lucida*, another of the many cycads endemic to Australia, the two sexes of the species have evolved some odorous tricks to manipulate pollinators into doing their bidding in a process known as push–pull pollination. In this species, each male cycad plant produces only one pollen cone. When its pollen is mature, the cone releases moderate quantities of a chemical called beta-myrcene, which is hugely attractive to small insects called thrips. The thrips come to feed on the pollen and crawl all over the cone, covering themselves with pollen during their feeding frenzy.

At midday, however, when daytime temperatures begin to rise, the heat increases the amount and potency of beta-myrcene in the pollen cones until the odor becomes an overpowering, toxic stench that is no longer attractive to insects, or to humans for that matter, as I know from experience in Thailand. The thrips, with their bodies heavily dusted with pollen, leave the pollen cone and seek out nearby female cones that are emitting lower, still pleasant amounts of beta-myrcene. When the thrips enter a female cone, hoping to feed on pollen again, they are fooled, finding only inedible ovules. Nevertheless, in their search for pollen in the female cone, the pollen from the male cone falls off the thrips' bodies onto the ovules, thereby cross-fertilizing the cycad.

As mentioned in chapter 7, one cycad family does not contain its seeds in a woody cone. Instead, cycads of the family Cycadaceae, which includes the sago palm, have leaf-like megasporophylls with a lamina at the tip and a stem-like base. Attached to either side of the stem are the ovules (fig. 8.8). This foliage-like form of the ovule-bearing structures is thought to represent a rare "missing link" that has preserved

Figure 8.8. The leaf-like seed-bearing megasporophylls of *Cycas revoluta. Left*, the head of megasporophylls in a sago palm cultivated in Huntington Beach, California. Photo by author. *Right*, two megasporophylls separated out from another individual plant, with large orange seeds along the petiole below the fringed leaf lamina, which shows the evolutionary origin of cycad megasporophylls as leaves. Photo by Ted Delevoryas.

the morphological stage of an ancient primitive leaf with ovules on its way to evolving into a woody seed cone.

While it does seem that cycads are dinosaur-age plants that got stuck somewhere on the road to evolving into modern-day plants, in actuality the groups of cycads that exist today on earth are relatively young. Instead of modern generic lineages reaching back to the Age of Cycads during the Mesozoic, the living genera of cycads actually first showed up in the Miocene, only 12 million years ago. This burst of diversification, documented by the molecular biology of living cycads, resulted in the global appearance of the modern genera *Cycas*, *Encephalartos*, *Macrozamia*, *Zamia*, and *Ceratozamia*. The driving force be-

Figure 8.9. A fossil cycad leaf of *Pseudoctenis ensiformis* from the Upper Jurassic of the Antarctic Peninsula. This incomplete frond is ca. 28 cm (11 in) long. Photo by author.

hind this 5-million-year-long burst of cycad evolution was likely global climate change that expanded optimal cycad habitats in the subtropics and tropics. Afterwards, however, as the earth cooled down during the Pliocene and the Pleistocene, the rate of evolutionary change in cycads slowed down to nearly nil. The decline in diversification is believed to have been in response to an increase in global seasonality and a decrease in suitable cycad habitats.

While the cycads today may be genetically quite different from their ancient ancestors, they looked pretty much the same in the Mesozoic. Most commonly found in the fossil record are cycad leaves, such as *Pseudoctenis* fronds, which bear short leaflets on either side of the leaf midrib (fig. 8.9). Each leaflet has robust veins that run parallel to one another to the tip of the leaflet. Strangely, though, the leaves of

Mesozoic cycads tended to be rather short compared with the longer fronds of today's cycads.

If the nutritional value of Jurassic cycad foliage was anything like that of today's cycads, they were probably not a favorite food of the Mesozoic megaherbivores. Not only were the Jurassic cycad leaves short in size, offering little biomass for consumption, but they were likely full of toxins and low in calories. Yet, herbivorous dinosaurs may have taken in a mouthful of cycad leaves once in a while. Even if cycads were not targeted as fodder by the giant Mesozoic herbivores, we should be grateful that their low nutritional appeal, high toxic content, and stunted leaf size helped them to fly under the radar as a source of dinosaur food. This suite of unpalatable characters may have contributed to the longevity of individual plants and, indeed, of the cycads as a whole, enabling them to survive through the approximately last 300 million years.

The living cycads, which form a sizeable group of some 300 species, are not a forlorn, diminished group of a few sole survivors today. Nevertheless, in view of their simple, primitive appearance and long fossil record spanning hundreds of millions of years, they can proudly bear the moniker "living fossil." Sadly, though, the outlook for cycads' survival is not entirely rosy. Because of excessive collecting in the wild, most cycads are threatened or endangered or have even gone extinct in their native habitats.

FURTHER READING

Frazer, J. 2013. Are cycads social plants? *Scientific American*. https://blogs.scientificamerican.com/artful-amoeba/are-cycads-social-plants/.

Jones, D. L. 2002. *Cycads of the World: Ancient Plants in Today's Landscape*. 2nd ed. Washington, DC: Smithsonian Institution Press.

Nagalingum, N. S., C. R. Marshall, T. B. Quental, H. S. Rai, D. P. Little, and S. Mathews. 2011. Recent synchronous radiation of a living fossil. *Science* 334, no. 6057: 796–99.

Symes, C. T. 2018. Cycad seed dispersal—The importance of large frugivorous birds. *Biodiversity Observations* 9, no. 4: 1–9. Also available in blog form at http://thebdi.org/2018/07/06/cycad-seed-dispersal/.

Ginkgo biloba

CHAPTER 9

The Maidenhair Tree, *Ginkgo*

With its deciduous leaves that turn a delicious golden color in the fall, *Ginkgo* is a distinctive tree that is instantly recognized by people all over the world (fig. 9.1). Many will also know that this tree is the last surviving species of its kind, qualifying it as a living fossil. In fact, *Ginkgo* is probably the most beloved living fossil in the plant kingdom. Its popularity might be attributed to its long paleontological history, its hard-luck story as a sole survivor, its role as a hardy street tree in urban settings, or its use as food and medicine. However, probably the most attractive characters of a *Ginkgo* tree are its fan-shaped leaves and their luminous yellow color in the fall.

The only species of its kind on earth today, *Ginkgo biloba* has flabellate, or fan-shaped, leaves. The leaf blade may be gently divided into two lobes, accounting for the species name *biloba*. In these cases, the leaf is not deeply incised but merely displays a single shallow notch in its outermost margin as a hint for dividing the lamina into two. Just as often, though, ginkgo leaves are unlobed, especially those leaves growing on the peg-like short-shoot branches. In any case, it is common for

Figure 9.1. *Ginkgo biloba. Left,* a 60-year-old tree bearing leaves with golden autumnal coloration in front of the paleontology building at the University of Bonn, Germany. *Right,* close-up of leaves in fall, in which nearly all of the green pigments have been broken down to unmask the yellow pigments. Photos by author.

both the unlobed and the gently lobed leaves to have a softly undulating edge, as well as a distinctive pattern in the venation. The pattern consists of two veins that enter the base of the leaf blade and then fork several times so that the fan-shaped leaf becomes evenly covered with veins. However, these veins hardy ever come back together, or anastomose, to form a reticulum, after they have diverged. This type of forking vein pattern, called dichotomous venation, is relatively rare. It is known only in the leaves of *Ginkgo,* in a few sporing plants, and, notably, in some of the most primitive terrestrial plants in the Devonian (see chapter 3).

Ginkgo's leaf shape is also the source of its popular name, the maidenhair tree. Its flabellate leaves are reminiscent of the leaves of some species of the maidenhair fern, *Adiantum*, though the ginkgo leaves and fern pinnules differ widely in size. The common maidenhair fern, *Adiantum capillus-veneris*, for example, has small, fan-shaped pinnules with a gently scalloped margin.

Ginkgo leaves are normally green throughout most of the year. However, in the fall, the green pigment in the leaves is broken down by the maidenhair tree to extract and save the valuable parts of the chlorophyll molecule. This process unmasks the yellow-colored pigments that were present in the leaves the entire time. After the breakdown of the chlorophyll, a ginkgo tree retains an unusually high quantity of various types of carotenoids in its leaves. However, the luminosity of autumnal *Ginkgo* leaves comes not only from an abundance of yellow pigments but also from two fluorescent compounds that develop only during their yellowing. These fluorescent substances, collectively called ginkgofluor, act as optical brighteners to produce the extreme golden coloration of autumnal *Ginkgo* leaves. This glowing fluorescence is especially evident at twilight. The fluorescent compounds are water soluble, which may account for the washed-out, pale lemon-chiffon-yellow coloration of fallen ginkgo leaves observed after a rain.

Beyond the unusual shape of their leaves, maidenhair trees look fairly commonplace, consisting of a single, sturdy, main trunk with a crown of green leaves on top. Mature trees normally reach heights of 20–30 m (65 to nearly 100 ft), although giant ginkgo trees of 50 m (164 ft) have been reported. Younger trees tend to have a narrow or pyramidal crown, while more mature trees have more spreading branches and a rounded crown. Oddly, although *Ginkgo* is a gymnosperm, it generally resembles many angiosperm trees. *Ginkgo* does not have the growth architecture of gymnosperms such as conifers (evergreen and commonly conical like a Christmas tree), cycads (narrow or

Figure 9.2. Nearly spent male cones in catkins and newly emerging leaves of *Ginkgo biloba*. Photo by Sten Porse, Creative Commons Attribution-ShareAlike 3.0 Unported license.

squat trunks with a top tuft of stiff fronds), or gnetophytes (evergreen trees, shrubs, and lianas).

Ginkgo trees reach maturity at about 20 years, at which time they begin to assume their characteristic tree form. More importantly, however, they also begin to express themselves sexually. Each maidenhair tree is either male or female. In the spring, male trees produce tiny cones made of pollen sacs containing *Ginkgo* pollen, which are borne in catkin-like structures (fig. 9.2). The structures develop on short shoots on the branches before the tree leafs out.

The female trees, by contrast, produce pairs of ovules attached at the end of a stalk, although it is common for only one of them to reach maturity. Once fertilized, the ovule will become a seed, which consists of the embryo encased in a hard seed coat. Each seed is then covered by a fleshy tissue called sarcotesta. Initially, the sarcotesta is green, but in the fall it, like the leaves, turns a beautiful golden color (fig. 9.3). In common parlance, these fleshy reproductive structures are known as ginkgo berries, while the seeds within are called ginkgo nuts.

Figure 9.3. The leaves and "berries" of *Ginkgo* turning a beautiful golden color in the autumn, as they do every year in Bonn, Germany. This photo shows a female branch that was grafted onto a male trunk of the *Ginkgo* tree planted in the Bonn University Botanic Gardens in 1870. Photo by author.

The stalked, ripened ginkgo berries have a silky surface sheen and look like golden cherries; they are aptly known in Chinese as "silver apricots." However, once the berries have fallen from the tree and lain on the ground for a few days, two volatile compounds start developing in the sarcotesta to give the berries a sickening, rancid smell, making them less attractive to humans. For this reason, male trees are mostly preferred over female trees in plantings in urban settings. Because it is difficult to tell the sexes apart for the first 20 years of their lives, a special variety of the *G. biloba* species called the "autumn gold cultivar" has been developed that produces only male trees.

The rancid smell of the berries is not necessarily a cruel trick played by the maidenhair tree on humans who unwittingly end up planting female *Ginkgo* trees. Instead, the smelly odor may serve as a clever ploy to get ginkgo seeds distributed by other mammals. It seems that some carnivores associate the odoriferous berries with the smell of rotting flesh. In Japan, for example, it has been observed that the raccoon dog and the masked palm civet swallow entire ginkgo berries. Later, their feces contain an abundance of ginkgo seeds that will then take the opportunity to sprout in the nutrient-rich carnivore droppings.

There is, however, one part of the female reproductive organ of ginkgoes that is popular with humans. In Asia, the ginkgo nuts are a culinary treat. After the nuts are collected and shelled, the nut meat inside must be boiled, stir-fried, roasted, or otherwise cooked because they are toxic when raw. Although ginkgo nuts are considered a great delicacy, whether sweet or savory, they are served in small amounts and usually with other ingredients so that they will not be eaten in great quantities (fig. 9.4). Extracts from ginkgo leaves or nuts are also used for medicinal purposes. The flavonoids are said to have powerful antioxidant qualities, while the terpenoids supposedly improve circulation by dilating blood vessels, thereby reducing the "stickiness" of platelets. These claims have not been scientifically proven, however.

Maidenhair trees are among the longest-lived trees on earth. Individual trees possess extraordinary longevity, reaching ages from hundreds to thousands of years. At the Royal Botanic Gardens, Kew, in London, the venerable old ginkgo tree planted in 1762 is affectionately called the Old Lion. It is still going strong, some 260 years later. In the Bonn University Botanic Gardens in Germany, a maidenhair tree planted in 1870 is revered not only for its uniqueness in the plant world and its longevity but also because it is bisexual. This unusual tree bears ginkgo berries and pollen catkins at the same time. It turns out that one of the favorite occupations of botanists in the nineteenth century was to graft branches from different sexes of the same species onto

Figure 9.4. A dessert of ginkgo nuts and strips of coconut in a light syrup served in Bangkok, Thailand. Photo by David Richfield, Creative Commons Attribution-ShareAlike 4.0 International license.

a parent plant. In Bonn, each main branch in the crown of the spreading ginkgo tree is marked with a small sign bearing either a ♀ or a ♂ symbol so that any perplexed plant enthusiastic can visually trace the source of sexual expression in the tree.

Ginkgo trees in the West do not compare in size or age to those in East Asia. Ginkgoes have long been revered in Buddhism, Taoism, Confucianism, and, in Japan, Shintoism. According to Shinto belief, maidenhair trees serve as a sacred place where local spirits dwell, which accounts for their profuse cultivation around temples. Huge trees with great girths are found near temples all over China, Japan, and Korea and were likely planted by monks. Some of these ginkgoes are locally reputed to be hundreds, if not thousands, of years old. One of the most accessible places for urban tourists to see huge ginkgoes is at the Buddhist temple complex in Asakusa in Tokyo, a famous site of worship, pilgrimages, and cultural festivals. Here, surrounding the Sensō-ji Temple, the oldest temple in Japan's capital city, there are ginkgo trees of immense size.

Some of giant trees at the Sensō-ji Temple produce strange, downward-growing woody shoots from their branches. Technically known as lignotubers, when these stalactite-like shoots come in contact with the soil, they take root and readily produce a clone of the parent tree (fig. 9.5). The clonal offspring plants can then form their own grove of ginkgoes and aid in the propagation of the genetic material of the parent plant. Thus, a ginkgo tree can reproduce asexually through its lignotubers, as well as sexually through the production of ovules or pollen.

In addition to vegetative reproduction, a number of innate characteristics, including their resistance to diseases and insect damage, play a role in ginkgo trees' longevity. *Ginkgo biloba* also grows in a wide spectrum of environmental conditions, ranging from temperate to subtropical climates and from forests to urban settings. Ginkgoes are popular street trees because of their great tolerance of air pollution. It is estimated, for instance, that 10 percent of all street trees in New York City are ginkgoes.

Recent physiological studies have shown that maidenhair trees are hardly affected by aging. The meristematic zones of growth, photosynthetic efficiency, seed germination rates, and gene expression of 600-year-old trees are just as vigorous as those of 15-year-old trees. The ability to regenerate and rejuvenate themselves certainly must have helped the ginkgo trees growing in Hiroshima survive and revive after the atomic bomb was dropped on that city in August 1945. These so-called A-bombed trees were able to come back to life, when so much was destroyed by the bomb's annihilating blast and then exposed to intense nuclear radiation. Six maidenhair trees within 2 km (1.2 mi) of ground zero, as well as 154 A-bombed trees of other species, live on today, serving as an outdoor green museum dedicated to peace.

Does individual longevity lead to species survival and evolutionary persistence through geological time? While this is a topic of current

Figure 9.5. Emerging lignotubers growing downward from branches of a *Ginkgo* tree. Photo by Maripo Goda, Creative Commons Attribution-ShareAlike 3.0 Unported license.

debate and quite pertinent to the concept of living fossils, it is difficult to say in the case of *Ginkgo*. However, the first ginkgophytes are believed to have appeared in the Permian, some 290 million years ago. Ginkgophytes then increased in biodiversity throughout the Mesozoic. The genus *Ginkgo* itself first appeared in the Jurassic, continued well into the Cenozoic, and survives to the present day.

In the fossil record, ginkgophytes are commonly identified through their leaves, although their seed-bearing structures and wood have also

Figure 9.6. The deeply incised leaf of a 174-million-year-old ginkgophyte, *Baiera gracilis*, from the well-known Middle Jurassic flora of Yorkshire, England. Photo by Georg Oleschinski.

been found. Like living *Ginkgo* leaves, fossil *Ginkgo* leaves had a fan-shaped lamina (see fig. 1.1). In the Mesozoic, however, the leaves of other kinds of ginkgophytes had lobes of various sizes. Sometimes they were so deeply incised that the lobes may appear today as separate segments of the same leaf (fig. 9.6).

The diversity of ginkgophytes, usually reflected in differences in leaf shape, size, and lobing, was much greater in the past. For example, several kinds of ginkgophytes grew in the Upper Jurassic Morrison Formation of North America, which is known for its dinosaur fauna. In southern Utah, at a fossil plant site appropriately named the Jurassic Salad Bar, at least four types of ginkgophyte trees grew along the shores of a freshwater pond and shed their leaves into it. One leaf type, known as *Ginkgoites*, resembles fossil ginkgophyte leaves known from elsewhere in the world, with a deeply lobed, fan-shaped lamina and diverging veins (fig. 9.7, left).

Another leaf type from the Jurassic Salad Bar, *Czekanowskia*, was a bit more unusual (fig. 9.7, right). The leaf of *Czekanowskia* was divided into long, narrow segments that occasionally divided again into a Y-shape. The narrow segments were fused together at the base of the leaf, reminiscent of a bundle of pine needles. However, in *Czekanows-*

Fig. 9.7. Two examples of 150-million-year-old ginkgophyte leaves from the Jurassic Salad Bar flora in southern Utah. *Left*, the deeply lobed leaf of *Ginkgoites cascandensis*. *Right*, the extremely incised lamina of a single *Czekanowskia* leaf. Photos by John Foster and Tom Howells.

kia, these segments did not consist of several needles, but were parts of a single leaf. Despite its extremely incised lamina, a *Czekanowskia* leaf had the fan shape so unique to maidenhair tree leaves (and some ferns) today. It also exhibited a forking pattern in its leaf segments and venation. The genus *Czekanowskia* went extinct at the end of the Mesozoic, but its relationship to ginkgoes is also confirmed by the chemical signature of its cuticle, the waxy layer on the outer surface of leaves. The cuticle chemistry of *Czekanowskia* is similar to that of living *Ginkgo biloba*. It is noteworthy that the habitat of the Jurassic Salad Bar ginkgophytes is the same sort of environment of disturbed streamsides and levees preferred by Cretaceous and early Cenozoic ginkgoes. In this way, the ecological preference of ancient ginkgoes differed from the native forested habitat of *Ginkgo biloba* in China today.

Finding fossil plants that shed light on the evolution of an important lineage always seems like such a stroke of excellent luck. In the case of the genus *Ginkgo*, a series of serendipitous discoveries in northeastern China and North American have truly led to a much deeper understanding of this plant's paleontological history. The preservation of 170-million-year-old *Ginkgo* berries and leaves from the Jurassic in China shows that there was an evolution in the seed-bearing organs from several berries in a cluster to today's single berry. The fossil record also shows a trend in the leaf shape from deep lobing in ancient forms to the single gentle notch in the living species. In the Cretaceous, 130 million years ago, the *Ginkgo* lineage in China split into two lines. One species retained the more primitive, Jurassic morphology of seed-bearing organs and leaves, while two other species evolved intermediate characters that are more similar to those of the Cenozoic species in North America. These latter forms then developed into those in our modern species of *G. biloba*. Thus, the Lower Cretaceous turns out to have been a pivotal time in the evolution of *Ginkgo*, with experimentation and diversification, which may have been triggered in part by the takeover by the up-and-coming flowering plants (see chapter 14) of the ginkgoes' favored ecological niche in disturbed stream habitats.

Later, in the Cenozoic, some 15 million years ago, ginkgo trees grew in a highly diverse subtropical forest during one of the warmer intervals of the Miocene. It may be hard to imagine it in today's cool, rainy Pacific Northwest. However, the remains of *Ginkgo*, along with remains of 34 angiosperm and 6 conifer species, were found as fossil logs and colorful pieces of silicified wood in basalt lava flows at the Ginkgo Petrified Forest State Park near Vantage, Washington State (fig. 9.8). Traces of minerals incorporated into the normally colorless silica during the fossilization of the wood produced a spectrum of colors including red to pink, all shades of brown, and black and white.

The story of *Ginkgo* is not without some more troubled times, however. Increasing global cooling during the Pleistocene seems to have caused near extinction. The genus became restricted to just one species, and its native distribution was limited to China. Recent genomic studies of living maidenhair trees growing in China and elsewhere in the world show that *Ginkgo biloba* took refuge in three places, in eastern, southwestern, and southern China, where they survived the Pleistocene glaciation cycles. In these refugia, they sheltered with other trees that are found today in single-species genera and do not occur naturally elsewhere in the world. Such uniquely monotypic plants include *Cathaya argyrophylla*, *Davidia involucrata*, and *Metasequoia glyptostroboides* (see chapter 12). From these relict populations in China, ginkgoes were then spread to other continents by humans in recent times and are widely planted today as ornamental trees.

It is actually no wonder that *Ginkgo* is considered the ultimate living fossil in the plant kingdom; it meets nearly all the criteria in Schopf's list (see chapter 1). The genus *Ginkgo* has survived more than 200 million years, with only small morphological changes. These changes include changes in leaf shape and the number of ginkgo berries borne on each stalk, which separate out taxa on the species level. The forked venation in ginkgo leaves reflects primitive morphological characters observed in some Devonian land plants, while the fan shape of the foliage is only found in a very few spore-bearing ferns

Figure 9.8. Pieces of silicified wood collected outside the Ginkgo Petrified Forest State Park in Washington State. Rounded and polished in a rock tumbler, they exhibit a multitude of fossil wood colors. Photo by Georg Oleschinski.

today. All evidence points to a refugium of the last surviving genus, *G. biloba*, in China, underlining the present low taxonomic diversity and relict distribution of the once widespread ginkgophytes in the Mesozoic. Thus, the extant maidenhair tree, *Ginkgo biloba*, is a clearly a morphological look-alike, a tenacious survivor, and a relict member of an ancient lineage and is indeed the quintessential living fossil.

FURTHER READING

Crane, P. 2013. *Ginkgo: The Tree That Time Forgot.* New Haven, CT: Yale University Press.

Del Tredici, P. 2008. Wake up and smell the ginkgos. *Arnoldia* 66, no. 2: 11–21.

Hori, T., R. W. Ridge, W. Tulecke, P. Del Tredici, J. Trémouillaux-Guiller, and H. Tobe, eds. 1997. *Ginkgo biloba: A Global Treasure; From Biology to Medicine.* Tokyo: Springer-Verlag.

Kwant, C. 1999. A-bombed Ginkgo trees in Hiroshima, Japan. *The Ginkgo Pages.* https://kwanten.home.xs4all.nl/hiroshima.htm. Photographs and documentary film.

Matile, P., B. M. P. Flach, and B. M. Eller. 1992. Autumn leaves of *Ginkgo biloba* L.: Optical properties, pigments and optical brighteners. *Botanica Acta* 105:13–17.

Royer, D. L., L. H. Hickey, and S. L. Wing. 2003. Ecological conservatism in the "living fossil" *Ginkgo. Paleobiology* 29:84–104.

Zhao, Y.-P., G. Fan, P.-P. Yin, S. Sun, N. Li, X. Hong, G. Hu, H. Zhang, F.-M. Zhang, J.-D. Han, Y.-J. Hao, Q. Xu, X. Yang, W. Xia, W. Chen, H.-Y. Lin, R. Zhang, J. Chen, X.-M. Zheng, S. M.-Y. Lee, J. Lee, K. Uehara, J. Wang, H. Yang, C.-X. Fu, X. Liu, X. Xu, and S. Ge. 2019. Resequencing 545 ginkgo genomes across the world reveals the evolutionary history of the living fossil. *Nature Communications* 10:4201.

Araucaria goroensis

CHAPTER 10

Araucarias, Kauris, and the Wollemi Pine

Thanks to the global ubiquity of Christmas trees and pine cones, most everyone is familiar with the boreal forests of firs and pines, while few are acquainted with the conifers of the Southern Hemisphere. In its fauna and flora, the austral biota differs from that of the Northern Hemisphere in featuring marsupials instead of placental mammals, araucarias and podocarps instead of firs, spruces, and pines. In some ways, these southern animals and plants may seem familiar to those in the north, but they can be unusual in their morphology, distinctive in their evolutionary history, and quite interesting in their own right. One prominent group of southern conifers is the family Araucariaceae, which embraces the sister genera *Araucaria*, *Agathis*, and *Wollemia*.

One of the oddest and most painful-looking conifers is *Araucaria araucana*, a native of Chile and Argentina. This evergreen tree has stiff, spine-tipped leaves that not only look prickly (fig. 10.1) but are truly painful to the touch. The person who coined its common name, monkey puzzle tree, must have done it tongue-in-cheek, for no self-

Figure 10.1. Branches of *Araucaria araucana* bearing prickly triangular leaves tipped with spring-green pollen cones at the Bonn University Botanic Gardens, Germany. Photo by Georg Oleschinski.

respecting primate would consider monkeying around on these branches. The monkey puzzle tree does have some redeeming characters, though. When the tree is young, it is highly prized as an ornamental. The beauty of the tree lies in its regular branching in whorls, or rings, around the trunk, which gives it an unusual architectural appearance. As the young tree becomes bushier and fills in with more branches, it can assume a pleasing, rounded shape (fig. 10.2).

Araucarias can attain heights of up to 80 m (262 ft), although most reach about 30–40 m (98–131 ft). They are long-lived and often tower over most other large trees that form the forest canopy. They are thus known ecologically as canopy emergents. In Brazil, however, mature trees of the native *Araucaria* species, *A. angustifolia* (fig. 10.3), range in height from 20 to 44 m (72 to 144 ft), nowhere near the maximum stature for the genus. *A. angustifolia* does grow more massively in girth, though. It is difficult to imagine a tree trunk more than 3.25 m (10 ft)

Figure 10.2. *Left,* a young *Araucaria araucana* tree with the whorled branch architecture so characteristic of araucarias. *Right,* the slightly older tree has a bushier shape formed by many densely spaced branches. The branches in the upper half of the tree are tipped with yellowish pollen cones, which are shown in detail in Figure 10.1. The two trees grow at the Bonn University Botanic Gardens, Germany. Photos by Georg Oleschinski.

in diameter, but that is indeed the girth of the stoutest native araucarian tree in Brazil. This particular tree is not especially tall, however, as it tops out at just over 39 m (128 ft). Sadly, giant araucarias have become increasingly rare in the Atlantic Forest biome (Mata Atlântica) of southern Brazil, and there are only 13 trees left with a diameter over 2 m (6.5 ft). The demise of araucarias in their native habitats, not only in Brazil but all over the world, is a disaster.

Conversely, in urban settings, cultivating araucaria trees has become very popular. With their great heights and pleasing, symmetrically branching architecture, their presence in a cityscape can be imposing. In the Boyle Heights neighborhood of Los Angeles, a towering, dome-shaped tree of *Araucaria bidwillii* has become a living monu-

Figure 10.3. The upward, candelabra-like branching style of two trees of *Araucaria angustifolia* at moonrise in the Atlantic Forest biome in southeastern Brazil. Photo by Christine Grabatin.

Figure 10.4. *Araucaria* as an icon. *Left*, the towering tree of *Araucaria bidwillii* in the Boyle Heights neighborhood of East Los Angeles can be seen for miles. *Right*, the beloved El Pino featured in a scene from the 1993 movie *Blood In Blood Out* on a T-shirt. Photos by author.

ment to the people of East Los Angeles. Locally called El Pino (Spanish for "The Pine"), this magnificent tree was grown from seed about 80 years ago by a Japanese dentist who had been interred in a World War II concentration camp in California. Although its origin is linked to ignominious events in California history, El Pino has since been iconized as a symbol of home, even evoking a sense of place in the community in its starring role in the 1993 Los Angeles cult movie *Blood In Blood Out* (fig. 10.4).

The branching architecture of *Araucaria* is distinctive and can be recognized from great distances, whether in a city or across a natural landscape. One characteristic is the regular symmetry of the main branches and the way they spread widely from the main trunk. As individual trees age, the branches on the upper part of the tree start reaching upward, while those on the lower trunk drop off completely. Mature individuals commonly resemble an upside-down umbrella or a candelabra, with the crown formed by long branches curving upward. The outer branches are the longest, while the central ones are the shortest. This contrasts with the branches of what most of us think of as a typical conifer, such as a fir tree, which has the downward, curved lines of a Christmas *Tannenbaum*. Some botanists have speculated that the unusual tree architecture of upswept branches in araucarias was an adaptation to protect the growth tissues at the branch tips from large herbivores (think of the huge plant-eating sauropod dinosaurs), but this hypothesis remains conjecture.

Species of *Araucaria* trees are variable in their sexuality, either bearing cones of both sexes on the same tree or bearing cones of only one sex on each tree. *Araucaria araucana*, the monkey puzzle tree with the extremely prickly leaves, for example, bears both pollen and seed cones on the same plant. While the araucarian pollen cones are not very flashy, the seed cones can be overwhelmingly large, woody, and heavy. The seed cones of *Araucaria bidwillii* (fig. 10.5), known as the bunya pine in its native Australia, can each reach a remarkable weight of 10 kg (22 lb) and a length of 30 cm (12 in).

Having such massive seed cones would not have adverse consequences if the cones stayed on the tree. However, many araucarias, such as the bunya pine, have what are called "shatter cones." Once mature, these heavy seed cones fall off the tree, hit the ground at a high velocity, and then shatter into pieces, casting numerous cone scales and their seeds in all directions. This successful dispersal mechanism has been used by araucarias since the Middle Jurassic; scattered cone scales containing the seeds of shattered cones have been found in the fossil record of the past 170 million years.

Figure 10.5. A massive, heavy seed cone of *Araucaria bidwillii*, here still green and relatively moderate in size at 22 cm (8.7 in). Note the big drop of white resin on the upper left of the cone. Photo by author.

As noted above, *Araucaria* is one of three genera in the family Araucariaceae, along with *Agathis* and *Wollemia*. Mature *Agathis* trees are commonly tall and have massive trunks. Their leaves are not needles but are much broader, more like the leaves of a flowering plant (fig. 10.6). *Wollemia* trees, on the other hand, have long, flat needles borne in four rows along each branch, as well as warty, chocolate brown bark.

Figure 10.6. *Agathis robusta*, the Queensland kauri, cultivated here at the Huntington Botanical Gardens, San Marino, California. *Left*, a 134-year-old tree measuring 34 m (112 ft) high with a diameter of nearly 600 cm (235 in). Because of its size it has been designated a California Big Tree. *Right*, a closeup of *A. robusta* leaves, which are wide and flat, not needle-like as in many conifers. Photos by author.

Although the genus *Araucaria* was described at the end of the eighteenth century, followed by *Agathis* at the beginning of the nineteenth century, it was not until 200 years later that *Wollemia* was discovered and took its rightful place alongside its sister genera. In 1994, David Noble, an Australian National Parks and Wildlife Service ranger, was hiking in his free time with two others when they happened upon some unusual trees. The trees topped out at 40 m (131 ft) and measured about 1 m (more than 3 ft) in diameter. They bore dark green needles in multiple rows on the branches and had dark brown bark that looked like bubbly chocolate.

Noble took a fallen branch back with him and showed it to Wyn Jones, of the New South Wales National Parks and Wildlife Service, and Jan Allen, of the Mount Tomah Botanic Garden. Later, the three of them returned to the site to collect more specimens for study. None of them had seen this sort of tree before. Together with another colleague, in 1995 Jones and Allen described it as a new genus of the conifer family Araucariaceae on the basis of its leaves and cones. Because the new genus was found growing in Wollemi National Park, had an elegant stature, and had been brought to their attention by David Noble, they named it the Wollemi pine, *Wollemia nobilis.* So far, there is only one known living species of the genus *Wollemia*.

The most striking thing about *Wollemia* is that it remained hidden in plain view only 150 km northwest of Sydney for 80 million years. At first, about 40 trees of *Wollemia* were found tucked away in a steep, narrow gorge in a very wet, warm temperate rain forest. The trees were multistemmed with coppice branches at the base of the trunk and a connected root system. Like araucarias elsewhere in the world, the oldest trees of *Wollemia* are canopy emergents. In their Australian native habitat, they rise above the tops of the neighboring angiosperm trees and tower over an understory of ferns and tree ferns. This population of *Wollemia* expresses its vigor in a healthy range of size classes, half of them being mature and the other half juvenile. There is also abundant seed production, as well as seedlings. Since its original discovery, nearly one hundred mature trees of *Wollemia* have been found at three different sites in the area.

Genetic studies show that the trees are all closely related to one another. It is thought that this lack of genetic variation occurred evolutionarily over time as a result of the population of *Wollemia* becoming so reduced in number—down to perhaps only one or two individuals. Because of the low numbers of single trees and populations, as well as the genus's lack of genetic variation, the Wollemi pine is listed as endangered by the Australian government.

The location of the Wollemi pine in Wollemi National Park is kept secret, and a commercial propagation program was organized in 2005 to provide Wollemi pine trees to botanical gardens (fig. 10.7) and gardeners to discourage the illegal harvesting of the trees from the wild. However, unauthorized visitors still managed to find the core populations, trekking in a pathogenic water mold that infects the roots of *Wollemia*. Other threats to this living fossil include the natural danger of fire, which is ever-present on the Australian continent. As recently as the 2019–20 bushfire season, for example, the natural groves of the Wollemi pine had to be saved by specialist firefighters, who installed an irrigation system and also dropped retardant to keep the last surviving species of *Wollemia* from perishing.

Altogether, there are nearly 41 species of the Araucariaceae, most of which are today native to the Southern Hemisphere. The only errant species are a few *Agathis* that cross the equator in Southeast Asia. While there is only one species of *Wollemia*, the other two genera, *Araucaria* and *Agathis*, comprise nearly the same number of species in the family. *Araucaria* is native to South America (Argentina, Chile, Brazil, Uruguay, and Paraguay) and Australasia (Papua, New Guinea, New Caledonia, Norfolk Island, and eastern Australia). The bulk of *Araucaria* species are found on the island of New Caledonia in the southwestern Pacific Ocean. The relatively recent formation of the island and its nickel- and cobalt-rich soils catalyzed a Cenozoic radiation of 14 *Araucaria* species. Like 75 percent of the island's unique flora, these 14 species are endemic, native only to New Caledonia. The fourteenth species, *Araucaria goroensis*, was newly discovered and described in this biodiversity hotspot in 2017.

In contrast to its restricted distribution today, *Araucaria* species grew all over the globe during the Mesozoic. In fact, araucarias were the dominant trees of the world's forests, along with conifer trees of the Podocarpaceae (see chapter 11). The Mesozoic world was very different from today's world, not only because araucarias and other coni-

Figure 10.7. Two young trees of *Wollemia nobilis* growing in front of the National Museum of Natural History at the Jardin des Plantes in Paris. The small dark green balls at the ends of the upper branches of the taller tree are developing seed cones, while the brown, icicle-shaped structures dangling from the branches below are pollen cones. Photo by author.

Figure 10.8. Segments of 225-million-year-old logs on a pleasant summer's day at Petrified Forest National Park in Arizona. Photo by author.

fers ruled the vegetation and dinosaurs roamed the earth but also because the climate was warmer and more equable across the latitudes. Huge fossil tree trunks from the Triassic and Jurassic strata of both the Northern and Southern Hemispheres bear witness to these widespread forests. Many of the fossil trunks are thought to be related to the Araucariaceae on the basis of their wood structure. Petrified Forest

Figure 10.9. Fossil evidence of *Araucaria* from the Middle Jurassic at Cerro Cuadrado Petrified Forest in Argentina. *Left*, a giant, 160-million-year-old araucarian tree trunk in the field. Photo by author. *Right*, a perfectly preserved seed cone of *Araucaria mirabilis* found at the same locality. The cone, on display at the Houston Museum of Natural Science in Texas, measures ca. 8 cm (3 in) in length. Photo by Daderot, Creative Commons CC0 1.0 Universal Public Domain Dedication.

National Park, in northeastern Arizona, is home to the best-known assemblage of fossil logs (fig. 10.8). There are, however, equally impressive sites with giant logs elsewhere in the world. One such site is the Cerro Cuadrado Petrified Forest in Patagonia, southeastern Argentina. The famous 160-million-year-old cones of *Araucaria mirabilis* come from this same Middle Jurassic locality (fig. 10.9).

There were no polar ice caps in the Mesozoic, so that luxuriant forests grew from pole to pole. About 150 million years ago, during the Upper Jurassic, at what is today the tip of the Antarctic Peninsula there was a lush forest of araucarian and podocarpacean conifers with a biodiverse understory of ferns and tree ferns. The araucarias are represented by fossils of fallen branches with the scaly leaves and shed

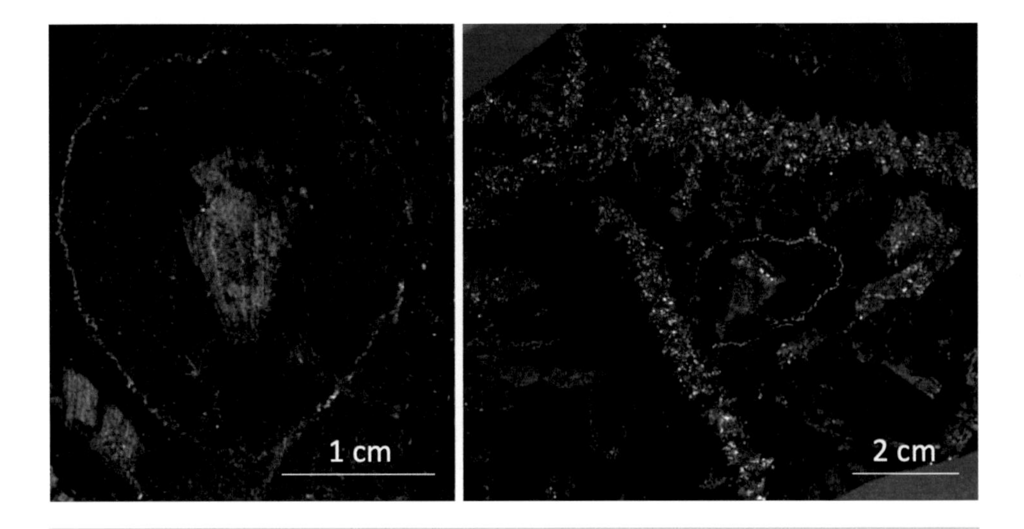

Figure 10.10. Fossil cone scales and branches of *Araucaria* from the Upper Jurassic Hope Bay flora at the tip of the Antarctic Peninsula. *Left*, a cone scale that has shed from an *Araucaria* shatter cone. *Right*, a detached *Araucaria* cone scale bracketed by branches presumably from the same tree. The branches are covered with scaly, pointed leaves. Photos by author.

seed scales of shatter cones (fig. 10.10), both of which are so typically found under *Araucaria* trees today.

Meanwhile, at the same geological time, the 150-million-year-old forests in western North America were also dominated by conifers, including several species of *Araucaria*. Some of these araucarian species are known in the rock record by their seed cones, others by their wood. When the fossil cones are cut lengthwise, they show the typical structure of *Araucaria* seed cones. They are characterized by a thick central core and cone scales each bearing only one seed, as observed inside a beautifully preserved cone from near Hanksville, Utah (fig. 10.11, left). In northeastern Utah, large tree trunks of *Araucaria* can be found near Vernal (fig. 10.11, center). When the wood of these fossil logs is sliced to a thickness of 30 micrometers (about half the thickness of a sheet of photocopy paper), microscope work shows that their anatomical structure is also typical of the wood of living araucarians. They possess copious amounts of resin that have solidified into resin plugs in the wood cells (fig. 10.11, right).

While *Araucaria* reached its diversity heyday in the Mesozoic and managed to survive throughout the Cenozoic, it is actively threatened with extinction today. As with the podocarps (chapter 11), the continual encroachment into tropical forests destroys the natural habitat of these magnificent trees. In particular, human activities on the mineral-rich island of New Caledonia are a specific worry for 11 of its 14 native species of *Araucaria*, as well as for the endemic species of *Agathis ovata* (fig. 10.12). Open-cast mining, fires, and habitat destruction endanger the existence of individual trees, the survivorship of the extremely rare

Figure 10.11. Araucarian seed cone and wood from the Upper Jurassic of Utah. *Left*, a longitudinally cut and polished 150-million-year-old cone from the Morrison Formation near Hanksville in southern Utah. Photo by Georg Oleschinski. *Center*, a fossil log from the Morrison Formation near Vernal. The log measures ca. 1 m (3.3 ft) across. Photo by author. *Right*, a wood thin section of the log in the center photo, cut in a radial plane of section. Note the dark brown resin plugs contained in the wood cells. Photo by author.

Figure 10.12. A seedling of *Agathis ovata* struggles to grow in the red-colored, nickel-rich soil of New Caledonia, the only place in the world where this species is native. Photo by author.

species, and the rich botanical diversity of the island. Once again, as mentioned in connection with the giant araucarias in Brazil, the menace of extinction to *Araucaria* is a major ecological concern across the globe today.

Nevertheless, despite the present-day pressure of extinction, the Araucariaceae are persistent survivors from the deep Mesozoic. They are robust, long-lived trees that have maintained their status as the tallest and oldest living beings in tropical forests in the Southern Hemisphere. As distinctive, beautifully majestic trees, they are also beloved by city dwellers and gardeners all over the world.

FURTHER READING

Enright, N. J., and R. S. Hill. 1995. *Ecology of the Southern Conifers.* Washington, DC: Smithsonian Institution Press.

Farjon, A. 2008. *A Natural History of Conifers*. Portland, OR: Timber Press.

Fassbender, T. 2022. El Pino Famoso: The majestic symbol of East Los Angeles. Los Angeles Explorers Guild. https://losangelesexplorersguild.com/2022/01/19/el-pino-famoso/.

Gee, C. T., H. M. Anderson, J. M. Anderson, S. R. Ash, D. J. Cantrill, J. H. A. van Konijnenburg-van Cittert, and V. Vajda. 2020. Postcards from the Mesozoic: Forest landscapes with giant flowering trees, enigmatic seed ferns, and other naked-seed plants. In *Nature through Time*, ed. E. Martinetto, E. Tschopp, and R. A. Gastaldo, 159–86. Berlin: Springer Verlag.

Larner, S. 2021. Bountiful bunyas: A charismatic tree with a fascinating history. State Library of Queensland. https://www.slq.qld.gov.au/blog/bountiful-bunyas-charismatic-tree-fascinating-history.

Mill, R. R., M. Ruhsam, P. I. Thomas, M. F. Gardner, and P. M. Hollingsworth. 2017. *Araucaria goroensis* (Araucariaceae), a new monkey puzzle from New Caledonia, and nomenclatural notes on *Araucaria muelleri*. *Edinburgh Journal of Botany* 74:123–29.

Podocarpus macrophyllus bonsai

CHAPTER 11

Podocarps

In many ways, the conifer family Podocarpaceae hides its candle under its hat. While taxonomically abundant in genera and species, the family is not as well known as its sister family Araucariaceae (chapter 10), at least not in the Northern Hemisphere. Yet, the podocarps are as fascinating as they are enigmatic. They are probably the strangest conifer family on the planet today owing to the wide variability in their growth habit, lifestyle, morphological traits, and ecological preferences. While the Podocarpaceae comprise 17 genera and 156 species, they also include several members that are endemic to eastern Australia and islands in the southwestern Pacific Ocean. A few oddball species have even pursued lifestyles that are rare among the conifers. Most podocarps, however, appear as conventional forest trees.

As with the southern conifer family Araucariaceae, the center of diversity of the Podocarpaceae lies today in the Southern Hemisphere. This hotspot of podocarp diversity is located in New Caledonia, Tasmania, New Zealand, and Malesia. Yet, native podocarps cross the equator northward into southern China, the Philippines, and southern Japan. The family also occurs in tropical montane Africa, as well as in the New World from Mexico to southern Chile.

In most of their range, podocarps assume the conventional growth habit of most conifers (fig. 11.1). They are tall, forest-forming trees with evergreen leaves. The tallest podocarp trees reach heights up to 25 m (82 ft) and are at home in tropical and subtropical forests. Like araucarias, podocarp trees help form the canopy when they attain their full height at maturity. When young, they contribute to the forest understory.

Across the family, however, the Podocarpaceae are highly variable in their growth habits and modes of life. They can take on the growth

Figure 11.1. Podocarpaceaous trees, old and young. *Left*, a mature tree of *Dacrydium cupressinum* growing in a montane forest on the South Island of New Zealand. Photo by Katja Schulz, Creative Commons Attribution-ShareAlike 2.0 Generic license. *Right*, a sapling of *Podocarpus macrophyllus* on a riverbank in a subtropical forest in China. As a young tree, it is now part of the understory; with time, however, it will achieve its place as a canopy tree in the forest. Photo by author.

Figure 11.2. The creeping pine, *Microcachrys tetragona. Left*, this podocarp is a sprawling shrub that grows among boulders in western Tasmania. Photo from Khan, Liu, and Biffin 2023, Creative Common CC BY license. *Right*, it is also known as the creeping strawberry pine because of its fleshy, bright red berries, as viewed here at Mount Field, Tasmania. Photo by Mnyberg, Creative Commons Attribution-ShareAlike 4.0 International license.

form of small, multistemmed trees, shrubs, or even low-growing, creeping trees. For example, *Microcachrys tetragona*, called the creeping pine or creeping strawberry pine, grows as a prostrate woody shrub (fig. 11.2). Restricted to mountain thickets and boulder fields in western Tasmania, *Microcachrys* bear tiny, edible red berries that at first glance are reminiscent of strawberry fruits. Such a creeping tree form is unusual among conifers.

In contrast to the creeping strawberry pine, the podocarp *Retrophyllum* is ordinary in its erect tree stance. Nearly all species of this genus form tall trees in habitats ranging from lowland and montane rain forests to high montane cloud forests. However, one species of *Retrophyllum, R. minus*, is a diminutive tree, reaching heights of only 2–3 meters (6–10 ft). Its small stature is likely owing to its penchant for growing in standing water (fig. 11.3). Found only on the southern end of New Caledonia, *R. minus* prefers the wet, flooded soils at the edges of lakes or grows directly in the water of slow-moving streams. Its

Figure 11.3. *Retrophyllum minus*, the New Caledonia corkwood, at the Madeleine Falls Nature Reserve in southern New Caledonia. This unusual podocarp is at home rooted in bodies of running water or on inundated shorelines. Photo by author.

common name, New Caledonia corkwood, refers to the corky, air-transmitting tissue that makes its trunk swell at its base. In its ecological preference for wetlands, the New Caledonia corkwood seems superficially similar to *Taxodium distichum*, the bald cypress of the bayous in the American South. However, while they are both conifers, the New Caledonia corkwood is a podocarp, while the bald cypress is a member of the cypress family.

If there were a contest for the oddest conifer, the prize would surely go to the podocarp *Parasitaxus usta*, the coral tree (fig. 11.4). While more than 3,000 species of parasites are known in the plant kingdom, the coral tree is the only parasitic conifer. Native to a remote, densely forested area of New Caledonia, *Parasitaxus usta* was first discovered and described in 1861. Yet, it was not until 1959 that it was definitively

Figure 11.4. The rare conifer parasite *Parasitaxus usta* in New Caledonia. Known as the coral tree, this parasite plant has deep wine to purple foliage and white seed cones. It is growing here on its podocarp host, *Falcatifolium taxoides*, which has the usual green leaves. Photo by naturewatchwidow at iNaturalist, Creative Commons Attribution-ShareAlike 4.0 International license.

determined to be a parasite. The leaves of the coral tree are not green but deep red to purple. These leaves are fleshy and scale-like. Strangely, this lack of green coloration is at odds with the amount of chlorophyll found in the coral tree. While the tree has abundant chlorophyll, many of the genes needed for photosynthesis are missing from the chloroplasts in *Parasitaxus usta*. Hence, this small tree must rely on another plant for its subsistence. The coral tree extracts nutrients from a species of a sister genus in the family, *Falcatifolium taxoides*, a podocarp that is also endemic to New Caledonia. *Parasitaxus* itself does not have roots

Figure 11.5. The versatility of *Podocarpus macrophyllus* allows it to be formed into a fanciful topiary in the Lotusland botanical garden near Santa Barbara, California. Photo by author.

but is always found growing attached to the roots or lower trunk of *Falcatifolium*. It is thought that *Parasitaxus* may actually feed on fungus, taking carbon from soil mycorrhizal fungi through its association with *Falcatifolium*.

The wide array of lifestyles and functional traits in the Podocarpaceae shows their genetic disposition to adaptation. Because of their morphological adaptability, in addition to their slow growth and dark green foliage, podocarps are horticulturally valuable. They are used in a wide range of cultivated settings in gardens and homes. Podocarps can be trimmed to form hedges or can stand alone as solitary, decorative trees, for example. One species of *Podocarpus*, *P. macrophyllus*, is called the Buddhist pine because it frequently grows as an ornamental around temples in Japan. This species is so versatile that the branches and foliage of large trees can be trained into topiaries (fig. 11.5) or the tree can be stunted in its growth and trained as a small bonsai.

Beyond podocarps' horticultural popularity, the forests of New Zealand are a stronghold of the trees in the natural world. There, as the dominant forest trees, podocarps form the basic framework of the plant community, which is filled in with lower-growing hardwood trees and an understory of shrubs and ferns. The podocarp family is represented by several taxa in the New Zealand forests—*Dacrycarpus*, *Prumnopitys ferrugineus*, *Prumnopitys taxifolia*, and *Podocarpus*—which also turn out to be ethnobotanically useful. For example, *Podocarpus totara*, locally known as the totara, is one the most sacred trees of the native Māori (fig. 11.6). Its wood is used for canoes, carvings, and furniture, while totara bark is collected for thatching roofs.

Living podocarps differ from other conifers such as araucarias and pines in not bearing their seeds in woody cones. Instead, they have evolutionarily reduced the number of cone scales in each cone from many to a few or even just one. During the growth of each seed cone, the infertile seed scales at the base of the cone fuse and become a brightly colored, berry-like fleshy tissue (fig. 11.7), while the uppermost scale develops into an ovule that becomes a seed when pollinated. The fleshy tissue at the base of the seed starts off green in color, becomes a cheery red or orange, then ripens to purple. Birds are attracted to the brightly colored tissue and inadvertently swallow the seed when targeting the fleshy tissue. The seed then passes unscathed through their digestive systems and is none the worse when dispersed through bird droppings.

Evolutionary studies suggest that ancient podocarps had small leaves and small, multiseeded cones in the Mesozoic. A reduction in cone size and in the number of seeds per cone occurred in the Lower

Figure 11.6. *Podocarpus totara*, a species native to New Zealand and sacred to the Māori, cultivated here at the Huntington Botanical Gardens in San Marino, California. Photo by author.

Cretaceous. Later, in the Upper Cretaceous, the family began evolving large leaves and large seeds. These trends in morphological changes were linked in paleontological time with the spread of closed-canopy forests and the diversification of small-sized birds. It is thought that the development of the single-seeded, fleshy, berry-like tissue marked the onset of the symbiosis between birds and the podocarps, an ecological swap of avian nutrition for podocarp seed dispersal that continues to this day.

In the Mesozoic record, the most common evidence of fossil podocarps is not, however, reproductive structures. Instead, twigs with leaves and sometimes pollen grains are the most frequently identified fossils of the family (fig. 11.8). Previously it was thought that the oldest podocarps hailed from the Triassic of the Southern Hemisphere, but

Figure 11.7. The reproductive structure of *Podocarpus chinensis*, the Chinese podocarp, which consists of a single powder-blue seed borne above a fleshy, berry-like tissue that ripens from green to red to purple. Photo by author.

recent discoveries have traced a deeper origin of podocarps back to the Permian in what is today Jordan. There, more than 250 million years ago podocarp trees thrived in lowland equatorial forests, along with seed ferns and bennettitaleans, two groups of gymnosperms now long extinct. Later, during the Jurassic and the Cretaceous, podocarps were able to expand their range globally. They codominated the world's forests with the Araucariaceae, as well as with Cheirolepidiaceae, an extinct conifer family. Whereas the Cheirolepidiaceae died out at the end of the Mesozoic, the Araucariaceae and Podocarpaceae soldiered on, marching toward the present day, although the global range of both families began to contract during the Upper Cretaceous and the Cenozoic.

As the family with the second-largest number of species among living conifers today, podocarps are certainly not sole survivors in the sense of a paucity of individuals or extremely low genetic diversity. However, several species survive today in extremely small patches, es-

Figure 11.8. Upper Jurassic *Elatocladus planus*, which is thought to represent leaves of the Podocarpaceae, from a climate warmer than that found today on the Antarctica Peninsula. *Left*, a branch or trunk wood beside fallen foliage. The distance between the 3 and 4 on the ruler equals 1 cm (ca. 0.4 in). *Right*, a single twig with long needles. Each tick mark on these rulers represents 1 mm. Photos by author.

pecially compared with the family's once larger biogeography. The genus *Microcachrys*, with its sole species, *Microcachrys tetragona*, the creeping strawberry pine, has been called one of the most spectacular cases of paleoendemism. This genus was once spread across the Southern Hemisphere during the Mesozoic and Cenozoic—in Antarctica, Australia, and New Zealand, on the now sunken islands in the middle of the Indian Ocean, and in southern Africa and South America. Today, however, *Microcachrys* is only known to grow in western Tasmania. This genus is readily identifiable in the fossil record by its pollen grains, which bear three air sacs in contrast to the usual one to two air sacs of other conifer pollen grains.

Sadly, too, the habitat destruction of tropical forests continually threatens surviving podocarp species with extinction. Despite its unique functional trait of parasitism among conifers, the coral tree is an example of a present-day species under ecological pressure from mining and quarrying on New Caledonia. Other menaces include activities triggered by humans such as an increase in wildfires and site disturbance caused by tourism.

Regardless of the threat of extinction and their reduced geographic dominance, the Podocarpaceae, like their sister family Araucariaceae, are tenacious survivors from the Paleozoic and the Mesozoic. As tough, slow-growing trees, they continued to hold their position as the tallest and oldest living beings in tropical forests in the Southern Hemisphere and beyond, despite the more than 250 million years of gradual and catastrophic biotic and climate change on earth. Today, both the Podocarpaceae and the Araucariaceae offer us an understanding of what conifer trees and global vegetation looked like during the reign of the mighty dinosaurs. Moreover, as the predominant species of the evergreen forest then, the Mesozoic podocarps and araucarias did more than serve as snacks for sauropods. They actively nourished, sustained, and fueled these megaherbivores to attain their gigantic size.

FURTHER READING

Blomenkemper, P., H. Kerp, A. A. Hamad, W. A. DiMichele, and B. Bomfleur. 2018. A hidden cradle of plant evolution in Permian tropical lowlands. *Science* 362:1414–16.

Feild, T. S, and T. J. Brodribb. 2005. A unique mode of parasitism in the conifer coral tree *Parasitaxus ustus* (Podocarpaceae). *Plant, Cell & Environment* 28:1316–25.

Khan, R., R. S. Hill, J. Liu, and E. Biffin. 2023.Diversity, distribution, systematics and conservation status of Podocarpaceae. *Plants* 12:1171.

Leslie, A. B. 2022. How many ways can you build a conifer cone? A commentary on "Origin and evolution of Podocarpaceae seed cones." *Annals of Botany* 130, no. 5: i–iii.

Leslie A. B., and R. B. J. Benson. 2022. Neontological and paleontological congruence in the evolution of Podocarpaceae (Coniferales) reproductive morphology. *Frontiers in Ecology and Evolution* 10:1058746.

Metasequoia glyptostroboides

CHAPTER 12

The Dawn Redwood, *Metasequoia*

Redwoods are known for their towering height, their longevity, and their reddish-brown bark and wood. Of the three genera of redwoods that remain today—*Sequoia*, *Sequoiadendron*, and *Metasequoia*—it is *Metasequoia* that deserves to be called a living being from the "Lost World." Also known as the dawn redwood, *Metasequoia* was described first as a fossil in Japan seven years before it suddenly appeared on the world stage as a conifer come back to life. The revelation of the dawn redwood as a living fossil is one of the most dramatic discovery stories of the twentieth century. It is also a tribute to a chain of dedicated botanists and to an intricate network of international plant science against the backdrop of the Second World War.

The story of *Metasequoia* begins in 1937, when a Japanese paleobotanist, Shigeru Miki, published a description of leafy twigs, small conifer cones, and seeds from the Pleistocene Osaka Group (fig. 12.1). These fossils were found in the Tokiguchi Porcelain Clay Formation on Honshu, the main island of Japan, which is now known to be Miocene in age, 12 million to 10 million years old. In his scientific paper,

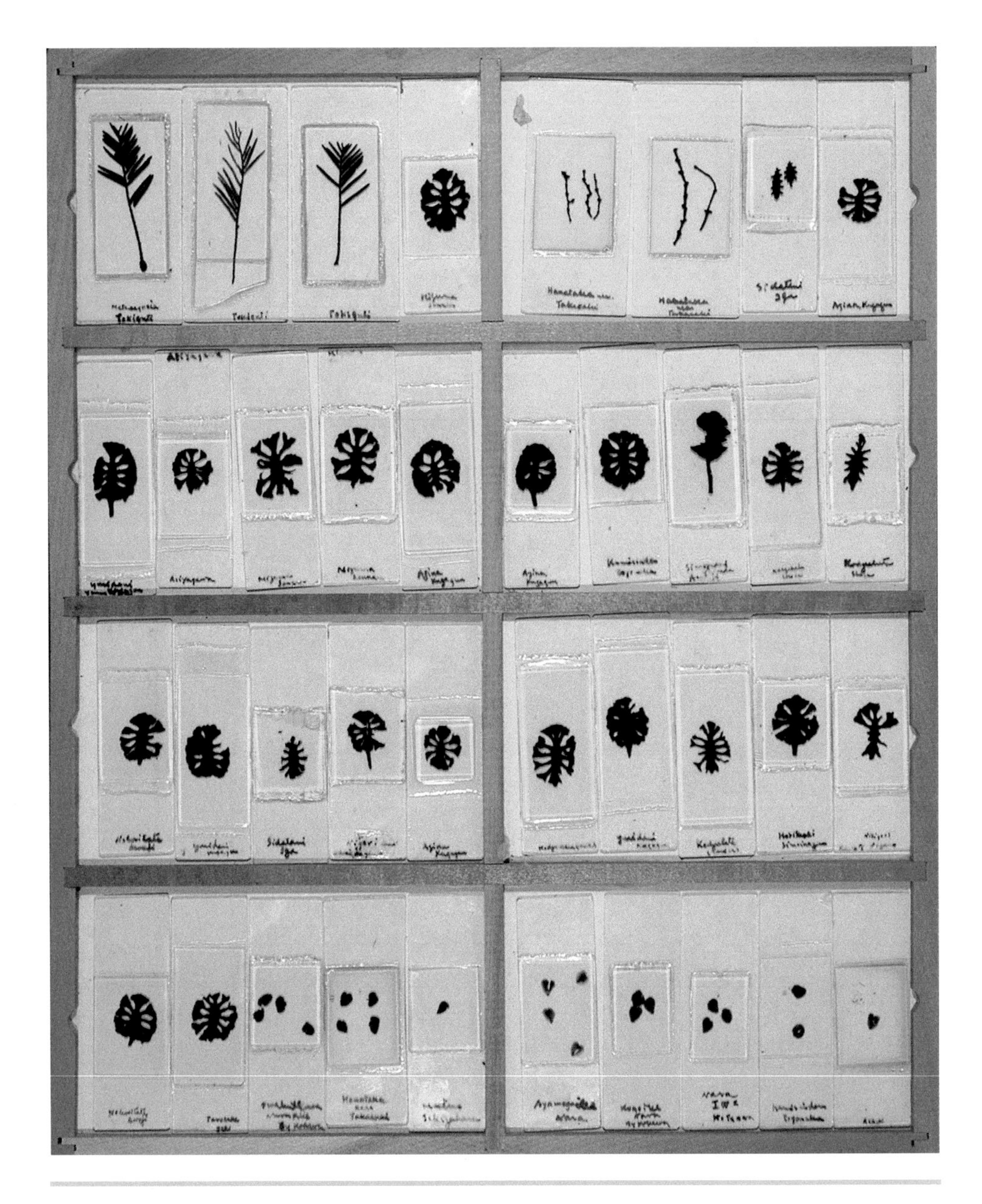

Figure 12.1. Glass slides with examples of the Miocene and Pliocene twigs, cones, and seeds described by Shigeru Miki in 1941 as a new genus and species, *Metasequoia disticha*. Photo courtesy of the Osaka Museum of Natural History, Japan.

Miki described these plant fossils as *Sequoia*, although he may have already suspected that they really represented something else. After two years of further deliberation, Miki must have gathered up his courage to announce his true thoughts. In 1941 he established an entirely new conifer genus, *Metasequoia,* for these fossils, one that was based on the long stalks of the seed cones and their opposite scales. The Greek-derived name of the new genus, in which *meta* means "after," certainly shows Miki's understanding that this new conifer was not *Sequoia* but transcended it.

The end of the 1930s and the 1940s were extremely turbulent years in Asia, the Second Sino-Japanese War having broken out in 1937. After the fall of the cities of Beijing, Tianjin, Shanghai, and Nanjing, the metropolis of Wuhan, on the banks of the Yangtze River, became the de facto wartime capital of the Republic of China, although the official capital was Chongqing, in Sichuan Province. Some 750 km separated Wuhan, in eastern Hubei Province, and the village of Moudao, where the *Metasequoia* tree was located, in western Hubei Province. It was an unstable time throughout China, with military invasions, refugees fleeing from the coastal cities, social upheaval, and bandits in the forests.

Despite the war and the state of confusion, plant lovers kept doing what they liked best—looking at plants and figuring out what they are. In 1943 Chan Wang, along with an assistant from the National Bureau of Forest Research, a division of the Ministry of Agriculture and Forests at Chongqing, went to Enshi, in western Hubei Province, to visit the Shennongjia forest area. On the way, they stopped off at the Wan Xian Agricultural High School, where Lung-Hsing Yang, the principal of the high school and Wang's classmate at Beijing University in the 1930s, asked him to identify a strange deciduous tree in Moudao, Lichuan, in Hubei Province. Wang obliged by altering his travel plans to detour by way of Moudao and collected leafy twigs and cones from the mystery tree on July 21, 1943. Once home from his trip, Wang deposited the material in the herbarium in Chongqing. Although he considered the tree to be nothing more than *Glyptostrobus pensilis*, called *shui song*, or "water pine," in Chinese, a species common to southeastern China, botanical science historians later identified the material collected by Wang as the first material collected of living *Metasequoia*.

The villagers at Moudao gave this tree a similar but definitely different name—*shui shan*, or "water fir"—because of its preference for moist sites with well-drained soils in ravines, on streambanks, or even rooted in standing water in swamps. The tree Wang collected from was a tall conifer measuring about 35 m (115 ft) in height and 2.4 m (nearly 8 ft) in diameter. In 2003 it was estimated to be around 400 years old.

In the summer of 1945, an assistant teacher at the National Central University in Nanjing, Chung-Lun Wu, went to Chongqing to study the conifer collection there. There, he came across Wang's herbarium material. Puzzled at Wang's identification of this conifer as *Glyptostrobus pensilis*, Wu, with Wang's permission, took one herbarium sheet with two cones to Wan-Chun Cheng, a dendrology professor at the National Central University, for further identification. Cheng realized that this conifer was not the water fir at all. Instead, the herbarium material represented a new genus owing to its opposite cone scales.

Intellectually engaged, Cheng sent his student Chi-Ju Hsueh to Moudao twice in 1946, in February and May, to make official collections of *Metasequoia*. Hsueh collected the new material from the same tree in Moudao that Wang had taken specimens from in 1943. In the fall of 1946 Cheng sent the herbarium specimens collected by Hsueh to a botany professor in Beijing, Hen Hsu Hu, asking him if he shared his opinion that the new mystery conifer was closely related to the coast redwood, *Sequoia*, and the giant sequoia, *Sequoiadendron*, in California. The realization that living *Metasequoia* (fig. 12.2) was a redwood was beginning to dawn on plant scientists.

This was the fortuitous moment in time when the chain of botanical and forestry links intersected with paleobotany. Professor Hu had gotten a reprint of Shigeru Miki's paper describing new fossil cones and needles in Japan from his assistant Shu-Hsia Fu. Hu recognized that the plant organs of the living plant were identical to those of Miki's new fossil genus *Metasequoia*. He also realized that the other reports

Figure 12.2. The dawn redwood, *Metasequoia glyptostroboides*, at the University of Bonn on a fine summer's day. This tree was planted by the paleobotany professor Hans-Joachim Schweitzer when the paleontology building was constructed in 1963. Photo by author.

of "*Sequoia chinensis*" in the fossil record of China were really reports of *Metasequoia*.

Cheng appealed to his plant science colleagues in the United States for funds for an expedition to collect more material of living *Metasequoia*. In June 1947 paleobotanist Ralph W. Chaney, at the University of California, Berkeley, sent $25. One month later, in July, botanist Elmer D. Merrill, at the Arnold Arboretum, Harvard University, sent a larger sum of $250. In the autumn of 1947, flush with travel funds, Cheng was able to send his assistant, Ching-Tsan Hwa, to collect seeds and reconnoiter the vegetation in the vicinity. Hwa arrived in Moudao in September and gathered seeds from the same tree of *Metasequoia* as Wang and Hsueh had. He also got a tip from the local inhabitants that there were many more trees of *Metasequoia* around the village of Shui-sha-pa, in what is now called Metasequoia Valley, where he went to continue his collecting activity. Hwa must have felt triumphant when he returned to Nanjing in November with about 1 kg (2.2 lb) of seeds collected from Moudao and Metasequoia Valley.

From Nanjing, *Metasequoia* seeds were sent off in late 1947, arriving at botanical gardens all over the world in early 1948. In May 1948 Hu and Cheng formally described living *Metasequoia* in a scientific journal. Three months after that, in August 1948, Hwa and Cheng returned to the living population of *Metasequoia* in Hubei Province to study its ecology. There, they took a moment off from their fieldwork to pose with a colleague in front of what was then the largest tree of *Metasequoia* in the area (see fig. 1.6). Sadly, this tree was killed in 1951 by a lightning strike.

Because Hu and Cheng had reached out across the Pacific—to Chaney at UC Berkeley and Merrill at Harvard—both the Chinese and the American botanists became instrumental in trumpeting the discovery of the dawn redwood, a new genus and species of conifer, as well as a living fossil. Chaney was so excited about the discovery of living *Metasequoia* that he even traveled to Metasequoia Valley from California with funding from the Save the Redwoods League to see the dawn redwood in person in 1948. He was accompanied by a science writer for the *San Francisco Chronicle*, Milton Silverman. Silverman chronicled their journey in great detail in a series of newspaper articles for the general public.

Thus, when the Second World War finally ended in the mid-1940s, the healing and rebuilding of the world could begin. The international heralding of the dawn redwood as a living fossil and the dissemination of *Metasequoia* seeds must have buoyed the world's spirits after the grueling and terrifying years of global war and spread a new sense of purpose among botanists to conserve and protect this precious living relict of the past. Today, many botanical gardens in the United States, Europe, and Asia have a small grove of dawn redwoods. All these groves are roughly three quarters of a century old, their birth year corresponding to the germination of seeds arriving from China in 1948.

Today, *Metasequoia* is represented by only a single species, *M. glyptostroboides*, which is native to one region in southeastern China. It grows naturally in the hills and wetlands of Lichuan County, including in the village Moudao, which is proudly referred to as "Metasequoia's hometown" by the local population. Relictual populations are found in species-diverse forests in Sichuan, Hubei, and Hunan Provinces. These areas receive moderate amounts of rainfall, averaging 1,300 mm (51 in) each year, and have an average annual temperature of 13°C (55°F). Yet, the seeds that were distributed all over the globe in 1948 showed that dawn redwood trees cultivated outside their native habitat could readily survive winter temperatures of −30°C to −40°C (−22°F to −40°F) in Canada and Russia. Clearly, this species can tolerate climatic conditions beyond those of its present-day native distribution.

Like *Ginkgo*, which is another of China's native trees, *Metasequoia glyptostroboides* is popular in its home country. Thoroughfares in many cities in China are planted with long rows of the dawn redwood. In the summer months, the linear arrays of the tall, straight trees give a regal, dignified look to the city streets. In the autumn the leaves turn rusty brown, and later in the winter the bare tree trunks point starkly toward the sky, resembling closely spaced sentries standing at attention (fig. 12.3). Dawn redwood trees grow tall and quickly, topping

Figure 12.3. Street plantings of dawn redwood trees in Shanghai. *Left*, trees with autumnally colored foliage in the fall. *Right*, trees without leaves in the winter. Photos courtesy of Biteena Frazier.

out at more than 40 m (131 ft) in height. Older trees have a trunk that tapers to the ground and a broad, buttressed base (fig. 12.4). The trunk can develop a burl, a rounded outgrowth filled with small knots of dormant buds, similar to those found on the coast redwood and the giant redwood.

Leaves of the dawn redwood are flattened needles arranged opposite one another in a flat splay. They are borne on the penultimate axes or occasionally appear as opposite pairs attached directly to the branchlet axis (fig. 12.5, left). In spring they sport a bright, fresh green color, and they turn reddish brown in the autumn before the branchlets fall off the tree. Fossil leaves and branches of *Metasequoia* look strikingly similar to recent foliage (fig. 12.5, right). The seed cones of living *Metasequoia* are relatively small, globoid to ovoid, and from 1.5 to 2.5 cm (up to 1 in) in diameter. They are similar in construction to those of the coast redwood, *Sequoia*, and the giant redwood, *Sequoiadendron*, but much smaller. Seed cones of the coast redwood are as long as 4 cm (about 1.5 in), while those of the giant redwood are even longer, up to 9 cm (3.5 in).

Figure 12.4. One of the first trees of *Metasequoia glyptostroboides* to be planted outside China, germinated from seed in 1948, in the Morris Arboretum in Philadelphia. Note the buttress roots. Photo by Katherine Wagner-Reiss, Creative Commons Attribution-ShareAlike 4.0 International license.

Figure 12.5. *Left,* a leafy twig from a living tree of *Metasequoia glyptostroboides.* Photo by Rasbak, Creative Commons Attribution-ShareAlike 3.0 Unported license. *Right,* a fossil leafy twig from a lower Paleocene species *Metasequoia occidentalis* from the Scollard Formation, central Alberta, Canada. Photo by georgialh, Creative Commons Attribution-ShareAlike 4.0 International license.

Deciduous conifers, ones that drop their leaves for the winter, are quite unusual. Today there are only five genera that annually shed their foliage: *Larix* (larch), *Pseudolarix* (golden larch), *Taxodium* (bald cypress), *Glyptostrobus* (Chinese swamp cypress), and *Metasequoia* (dawn redwood). Given that both *Metasequoia* and *Glyptostrobus* are deciduous and grow in standing water, it is no wonder that Hu and Cheng gave *Metasequoia* the specific epithet *glyptostroboides.* The leaves and seed cones of what the Chinese call the water fir and the water pine are quite different, though.

While the remarkable character of deciduousness found in *Metasequoia* is not shared with the other redwoods, *Sequoia* and *Sequoiadendron,* all three trees have the same distinctive reddish bark and wood that turns red when exposed to air. The red color is indicative of high levels of tannins, which help protect the trees from pathogens such as fungi and insects. Accordingly, wood of redwoods, especially of those in California, was used for many decades when a rot-resistant material was needed for building construction or outdoor furniture.

The three genera—*Metasequoia*, *Sequoia*, *Sequoiadendron*—are each represented by just one species, and these three species represent the end of their respective lineages. Morphological and genetic studies show that they cluster in their own subfamily, the Sequoioideae, within the globally distributed cypress family, the Cupressaceae. The chromosome count of the coast redwood, *Sequoia*, is extremely unusual because it is hexaploid (2n = 66) instead of the usual diploid (2n = 22). This polyploidy, plus further genetic analysis, suggests that *Sequoia* was the result of a hybridization event between *Metasequoia* and another close relative, either an extinct ancestor or possibly *Sequoiadendron*. Recently, 30-million-year-old *Metasequoia* leaves from West Siberia were observed to contain 22 chromosomes, exhibiting a karyotype similar to that of the modern species of dawn redwood.

In all cases, the present-day natural ranges of the redwoods are limited to very small areas on either side of the Pacific Ocean. While the dawn redwood has a patchy distribution in Sichuan, Hubei, and Hunan Provinces in China, the coast redwood thrives in the fog-filled forests along the coasts of northern California and southern Oregon (fig. 12.6, left). The giant redwood has the narrowest range, occurring naturally only on the western slopes of the Sierra Nevada in California (fig. 12.6, right).

While *Metasequoia* is a sole survivor today in the sense of being the last species and genus of its kind, it was actually a global player in the Northern Hemisphere 100 million years ago. Starting in the Upper Cretaceous and extending to the end of the Pleistocene, about 10,000 years ago, *Metasequoia* grew in the middle and high latitudes of North America and Eurasia. It was one of many species in broad-leaved forests and occurred in thick stands in lowland swamp forests, particularly in the polar regions.

Figure 12.6. *Left*, trees of the tall and narrow coast redwood, *Sequoia sempervirens*, in Jedediah Smith State Park, near Crescent City, California. Photo by Acroterion, Creative Commons Attribution-ShareAlike 3.0 Unported license. *Right*, the more massive form of *Sequoiadendron giganteum*, here the famous "Grizzly Giant" giant redwood tree of Mariposa Grove, Yosemite National Park, California. Photo by Mike Murphy, Creative Commons Attribution-ShareAlike 2.0 Generic license.

During the warmest epoch of the Cenozoic—the middle Eocene, roughly 45 million years ago—there were extensive forests with *Metasequoia* on Ellesmere Island and Axel Heiberg Island in the Canadian Arctic Archipelago. At these fossil sites, *Metasequoia* trunks are commonly found in original growth position, along with an abundance of fossil wood, needles, and cones. During the Paleocene and the Eocene, these sites were located at around 80°N. However, the climate at that latitude was very different from what it is today, for it was warm temperate then. As they do today, the plants had a problem for three months in the winter when there would have been complete darkness. Today, trees cope with this annual period of darkness by shedding their leaves and shutting down physiologically during the months-long polar night. It is thought that *Metasequoia* adapted this strategy early on, evolving from an evergreen tree to one that was deciduous, before it expanded its range into high latitudes, which allowed it to become a dominant tree in the far north during the mid-Cenozoic.

About 23 million years ago, starting in the Miocene, increasing global aridity, cooling, and competition from members of the pine family began to decrease the biogeographic range of *Metasequoia*. During the Pleistocene, *Metasequoia* could still be found in some limited parts of North America and Europe, but it was most common in East Asia. By the Holocene, some 10,000 years ago, however, its range was restricted to Japan and China, before it became limited to southeastern China, its native habitat today. Thanks to dedicated botanists and foresters 80 years ago, this water-loving conifer was recognized as unique, pulled out of oblivion, and rescued from possible extinction. Today, the dawn redwood is a popular tree that grows all over the world and is given its rightful due as a living fossil.

FURTHER READING

LePage, B. A., C. J. Williams, and H. Yang, eds. 2005. *The Geobiology and Ecology of Metasequoia*. Dordrecht: Springer.

Ma, J. 2002. The history of the discovery and initial seed dissemination of *Metasequoia glyptostroboides*, a "living fossil." *Aliso* 21:65–75. Includes a photo of the "type tree" in Moudao, Lichuan, Hubei, China.

———. 2003. The chronology of the "living fossil" *Metasequoia glyptostroboides* (Taxodiaceae): A review. *Harvard Papers in Botany* 8:9–18.

Ozerov, I. A., N. A. Zhinkina, A. A. Torshilova, E. M. Machs, and A. V. Rodionov. 2021. Chromosomes of fossilized *Metasequoia* from early Oligocene of Siberia. *Review of Palaeobotany and Palynology* 287:104365.

Williams, C. J., A. H. Johnson, B. A. LePage, D. R. Vann, and T. Sweda. 2003. Reconstruction of Tertiary *Metasequoia* forests. II. Structure, biomass, and productivity of Eocene floodplain forests in the Canadian Arctic. *Paleobiology* 29:271–92.

Sciadopitys verticillata

CHAPTER 13

The Japanese Umbrella Pine, *Sciadopitys*

On Honshu, the largest island of Japan, the Buddhist Jinguji Temple, located some 70 km (43.5 mi) north of Kyoto, near the city of Obama, has a conifer tree with a moderate height of 27 m (89 ft) but a massive girth of more than 4 m (13 ft). This conifer is a Japanese umbrella pine. Historical records show that it has been worshipped at this temple for the last 700 years, when a monk from a nearby temple planted the tree. Because the main Buddhist deity of the temple is female, women come to touch the giant umbrella pine at the annual festival in hopes of being blessed with healthy children. The umbrella pine of Jinguji Temple is so revered that it has been designated a natural monument.

The umbrella pine, or *kōyamaki*, is native to Japan and has played a major role in Japanese culture and religious rites for many centuries. In the Kyoto area of central Japan, for example, the *kōyamaki* has been associated with Buddhist temples and become part of Buddhist prayers. On Mount Kiso, in the alpine region of Nagano, *kōyamaki* branches

Figure 13.1. A typical cemetery in Japan, here in Kyoto, with neatly organized family plots, stone monuments, and several *kōyamaki* trees. Photo by author.

are placed on the graves of loved ones to help lead the spirits back to the land of the living. To this day, many Japanese grave sites are family plots with stone monuments that include a Japanese umbrella tree planted nearby (fig. 13.1) or are adorned by *kōyamaki* branches in a vase.

The Japanese umbrella pine is one of the sacred trees of the Kiso forest, in the mountains of the Nagano Prefecture. Kiso is home to five sacred trees, all of which are conifers native to Japan and highly valued for their wood. Shinto shrines, for example, are commonly built from the untreated wood of these sacred trees. The natural wood of *Sciadopitys* is soft, light-colored, spicy in scent, and fine-textured. These qualities, as well as the moisture-repelling qualities of the wood, have made it valuable not only for constructing important buildings but also for making boats and coffins.

It turns out that the Japanese umbrella pine, known botanically as *Sciadopitys verticillata*, is not actually a pine. Instead, it is a solitary genus and species in its own family, the Sciadopityaceae. This lone species used to be classified in the same family as *Taxodium*, the bald cypress. However, it is now known to have no close relatives among the conifers. In fact, some botanists isolate the umbrella pine even further, locating it in its own order, the Sciadopityales, although most put it in the Coniferales.

The genus name *Sciadopitys* comes from the Greek *skios* or *skiados*, "umbrella," and *pitys*, meaning "pine," reflecting the usual misnaming of conifers in general as pines. The tree's species name, *verticillata*, refers to the whorled arrangement of its leaves, which densely encircle the branches in rings. The common name Japanese umbrella pine adds even more confusion to the tree's botanical identity. This is because there is another tree known as the umbrella pine, *Pinus pinea*, which is a true member of the pine family and native to the Mediterranean (fig. 13.2, left). In these two unrelated trees, the word *umbrella* describes a shape in different parts of the plant. In the Mediterranean *Pinus pinea* it refers to the concave-down crown of the pine trees, while in the Japanese tree it is the clusters of needle-like structures on the branches that resemble the radiating ribs of a parasol (fig. 13.2, right).

Figure 13.2. Umbrella pines. *Left,* the Mediterranean umbrella pine, *Pinus pinea,* growing in Spain, with its characteristic umbrella-shaped crown. The widely globose seed cone of this species is featured in Figure 7.7. *Right,* a Japanese umbrella pine, *Sciadopitys verticillata,* planted at Ballard Locks Fish Ladder in Seattle, Washington. Its green, needle-like structures radiate in whorls like the ribs of small umbrellas. The seed cones of this species are smaller and egg-shaped (see also Figure 13.3). Photos by author.

An evergreen conifer, *Sciadopitys* takes on a pyramidal or narrowly conical shape after some years (see fig. 13.1). It is a slow grower but can reach heights of 15–27 m (50–88 ft). In gardens it is commonly used as an eye-catcher because of its striking pointed apex, reddish-brown bark, which peels off in vertical strips, and deep green foliage. The needle-like leaves can grow up to 13 cm (5 in) long. The seed cones are egg-shaped and 6–11 cm (about 2 to 4.5 in) in length, become mature in about 18 months, and are much less massively built than those of pines or araucarias (fig. 13.3).

Today, the natural distribution of *Sciadopitys* is restricted to western and central Japan. The trees are found in small patches on the Japanese islands of Honshu, Kyushu, and Shikoku. They grow naturally in mixed forests of conifers and angiosperms in the mountains at mid-altitudes from 500 to 1,000 m (1,640 to 3,281 ft) and in areas with high rainfall and humidity. In the latter habitat, it is remarkable how the low clouds and fog blanket the forest. The tiny biogeographic range of *Sciadopitys* in the present day may be a result of climate change during the glacial and interglacial periods of the last ice age, as well as deforestation. Pollen studies show that *kōyamaki* used to cover the Osaka Plain, but it is thought that its exploitation in building construction and especially in coffin making, particularly in AD 200–700, may have contributed to the decimation of its larger geographic range. In fact, the overharvesting of *Sciadopitys* wood in earlier periods of East Asian history may have started its demise. For instance, in the kingdom of Kudara, located in the southwest of the Korean peninsula, all the emperors' funerary caskets from 1800 to 660 BC were made of *Sciadopitys* wood. At that time, as well as now, the Japanese umbrella pine grew only in Japan. Finding its wood in Korea implies that it must have been especially commissioned and transported abroad for 1,200 years.

Fig. 13.3. A seed cone of *Sciadopitys verticillata* collected from a tree planted in Bonn, Germany. Note that the cone scales are relatively flat and have a flap-like edge. Oddly, this particular seed cone terminates in a pair of needles. The cone measures 8.5 cm (3.5 in) from the tip of the needles to the bottom of its stalk. Photo by author.

The clusters of leaf-like structures of *Sciadopitys* are distinctive and particularly striking in form owing to their whorled arrangement. Despite their appearance, these clusters are not actually made up of leaves. Instead, they are flattened green stems called cladodes, which function as the photosynthetic organs of the tree. The evolutionary origin of the cladodes from stems is confirmed by the presence of internal stem tissues. On the outside, unique characters of the cladodes include a so-called double-tipped apex at the end of the leaf, a central groove on the upper side, a deeper furrow on the lower side, and a bulbous base. The true leaves of *Sciadopitys* are found attached to the branch below the cladode and take the form of small brown bracts. Two vascular veins deep inside the cladode run along either side of the central groove. These veins and the central groove are also observed on fossil leaves of *Sciadopitys.*

As with all fossil plants, the distinctive features of the living Japanese umbrella pine are used to identify *Sciadopitys* in the fossil record. However, superficial appearances can be deceiving; similar morphological characters can arise several times in the evolutionary history of unrelated plant groups. Apparently, similar leaves with a median groove on their undersides were common in the Northern Hemisphere from the Lower Jurassic to the Lower Cretaceous, roughly 200 million to 100 million years ago. While these fossil leaves have been described as "*Sciadopitys*-like," most paleobotanists are in agreement that they most likely represent a different group of conifers. These fossil leaves are instead placed in an extinct family, the Miroviaceae.

Similarly, "*Sciadopitys*-like" wood has been reported from the Mesozoic from various sites around the world. The oldest unequivocal evidence of *Sciadopitys* in the Mesozoic comes from the Upper Cretaceous, about 95 million years ago. In Japan, well-preserved fossil seed cones, cone scales, seeds, wood, and cladodes of the family dating from the end of the Cretaceous to the Pliocene have been recorded from several localities.

In Europe, *Sciadopitys* played a very important role in the lowland forests in Cenozoic time. Two cladodes of *Sciadopitys*, complete with a double tip, a central groove on the upper side, a deeper furrow on the lower side, a slightly swollen leaf base, and other fine features, have been found embalmed in Baltic resin of Eocene age, from 47 million to 35 million years old. As apparent in its name, Baltic amber is fossil resin from the Baltic region. These deposits are famous for their wealth of entrapped insects, as well as of other animals and plants. Although it is known that the flora in Baltic amber represents a mixed conifer-angiosperm forest of mostly pines, cypresses, oaks, and laurels, it is noteworthy that the umbrella pine contributed to the high diversity of the lowland forest ecosystem during the highest temperatures and warmest year-round climate of the Cenozoic.

A bit later in time, the umbrella pine expanded its geographic and habitat range in Europe. It became the dominant tree in the lowland swamp forests along the Proto–North Sea in western Germany during Miocene times. Mapping of fossil stumps and identification of wood in a fossil forest in which the trees were preserved still rooted in living position show that the swamp forests were dominated by *Sciadopitys*. Today, the swamp forests are preserved as a thick seam of lignite coal (fig. 13.4, left). Fossil leaves of *Sciadopitys* can be found in the lignites, but the more common fossil is actually *Sciadopitys* wood (fig. 13.4, center and right). The wood from this 20-million-year-old forest is so well preserved that it is as fresh in appearance and texture as recent conifer wood. This degree of preservation typifies what is called "unaltered wood" in paleobotany. The defining features of the fossil *Sciadopitys* wood are the large, round openings in its ray cells. On this Miocene forest floor, *Sciadopitys* foliage and pollen were also found between the fossil stumps, confirming that these were indeed *Sciadopitys trees*. Other trees in the Miocene swamp forests were other kinds of conifers and, remarkably, giant palms with huge trunks more than 1 m (more than 40 in) in diameter. The peat deposit formed by the decomposing trees and carbon-rich plant litter eventually became the mighty lig-

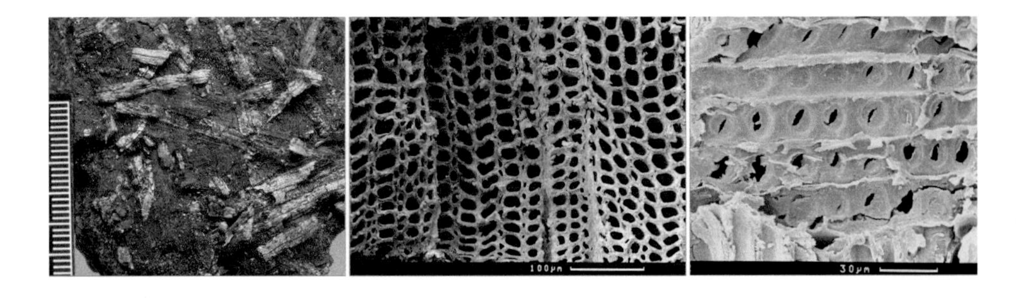

Figure 13.4. *Left*, 15-million-year-old leaves of *Sciadopitys* from the Garzweiler open-pit lignite mine near Bonn, Germany. A central groove can be observed on some leaves. Each line on the scale represents 1 mm. Photo by Georg Oleschinski. *Center*, a photo taken with a scanning electron microscope of *Sciadopitys* wood from the same locality. Photo by author. *Right*, a higher-resolution photo taken with a scanning electron microscope of the ray cells in the same *Sciadopitys* wood. Note the large, round openings, called crossfield pitting, so distinctive of *Sciadopitys* wood. Photo by author.

nite coal seams that powered the industrial regions of northwestern Germany in the nineteenth and twentieth centuries.

Midway through the Cenozoic era, the habitat tolerances of *Sciadopitys* must have shifted, perhaps in response to the global cooling that started at the end of the Eocene, 34 million years ago. After its high point in the Northern Hemisphere during the Eocene, Oligocene, and Miocene, the biogeographic range of the genus began to constrict. Whereas it was once the dominant tree of the coastal swamp forests in the early Cenozoic of Europe, the natural distribution of *Sciadopitys* is restricted today to Asia, where it grows in small patches in temperate mountain cloud forests in Japan. Nevertheless, despite an extreme reduction in its taxonomic diversity, the Japanese umbrella pine has continued to exist for more than 100 million years since the Mesozoic. It remains alive as the sole survivor of its kind and thus qualifies as a true living fossil.

A Botanical Pilgrimage

In 2012, when my family and I decided to tour central Japan for three weeks, it was my chance to see *Sciadopitys* in its native habitat. I knew the Japanese umbrella pine from my own potted plant collection, as well as from wood and pollen fossils from the Miocene lignites near Cologne and Bonn. However, I really wanted to see living trees in the wild. To this end, we recruited a colleague, Minoru Tsukagoshi, a paleobotanist at the Osaka Museum of Natural History, to take us to one of the last remaining refuges of *kōyamaki.* I knew from my research on the Miocene lignite forests that *Sciadopitys* grows today in the mountains. What I had not realized was that its habitat is a cool and humid temperate rain forest, reminiscent of the temperate rain forests in the Pacific Northwest of the United States.

Minoru rented a van big enough for the seven of us and drove us to Mount Odaigahara, southwest of the city of Nara. The Odaigahara Plateau is one of 100 mountains in Japan known for stunningly beautiful views, as well as for their spiritual and historical importance. In fact, this area is so important biologically that in 1980 an area of 36,000 hectares in the region of Mount Odaigahara (elevation of 1,695 m, or 5,561 ft) and nearby Mount Omine (elevation of 1,719 m, or 5,640 ft) was designated a UNESCO Biosphere Reserve. Mounts Odaigahara and Omine are home not only to a wide array of plant communities but also to a biodiverse mammal fauna that includes the Asian black bear, the sika deer, and many species of birds.

From the trailhead the walk to our destination was 4.4 km (2.7 mi) uphill, which took us just over an hour. As the trail made its way upslope, the level of the clouds descended, and we were soon enveloped in a thick fog. All of a sudden, a sika deer materialized out of the fog and, amazingly, posed for us calmly and unafraid for several minutes (fig. 13.5).

Soon afterwards, wet with cloud condensation, we reached the site of *Sciadopitys* at the top of the mountain. The Japanese umbrella pines were tall and magnificent, extending their branches in the damp fog.

Figure 13.5. A sika deer appeared magically in the mist of the cloud forest on our hike up to Mount Odaigahara, near Nara, Japan. Photo by author.

The trees did not have the perfect, groomed appearance of their garden relatives but were more varied in size and shape (fig. 13.6). It was clear from their missing branches and asymmetrical silhouettes that they lived under real-life conditions; they were survivors.

Here, in one of their last refugia, the Japanese umbrella pines are not just surviving but thriving. Seeing several mature trees bearing seed cones was gratifying. It was also heartwarming to see young robust seedlings holding the banner high for the new generation of *kōya-maki* trees (fig. 13.7).

Figure 13.6. The Japanese umbrella pine in its native habitat of temperate cloud forest on Mount Odaigahara, Japan. Photo by author.

Figure 13.7. The population of *Sciadopitys verticillata* trees on Mount Odaigahara is small, but their vigorous growth and the occurrence of seed cones and seedings suggest that the trees are thriving. Photo by author.

Our botanical pilgrimage to one of the last places on earth where natural stands of *Sciadopitys* still exist was completed (fig. 13.8). We were happy to see the Japanese umbrella pine flourishing in its chosen habitat today in the cool, damp, mid-altitude mountains of Japan. Our trek to Odaigahara was not only successful; meeting up with a plant survivor from the Mesozoic was unforgettable.

Figure 13.8. Mission accomplished for the botanical pilgrims to the *Sciadopitys* forest on Mount Odaigahara in Japan. *Left to right*: Philipp Sander, Chris Sander, Carole Gee, Martin Sander, Shoji Hayashi, and Laura Sander. Photo by Minoru Tsukagoshi.

FURTHER READING

Mosbrugger, V., C. T. Gee, G. Belz, and A. R. Ashraf. 1994. Three-dimensional reconstruction of an in situ Miocene peat forest from the Lower Rhine Embayment, NW Germany—New methods in palaeovegetation analysis. *Palaegeography, Palaeoclimatology, Palaeoecology* 110:295–317.

Sadowski, E.-M., A. R. Schmidt, L. Kunzmann, C. Gröhn, and L. J. Seyfullah. 2016. *Sciadopitys* cladodes from Eocene Baltic amber. *Botanical Journal of the Linnean Society* 180:258–68.

Saiki, K. 1992. A new sciadopityaceous seed cone from the Upper Cretaceous of Hokkaido, Japan. *American Journal of Botany* 70:989–95.

Tsukada, M. 1963. Umbrella pine, *Sciadopitys verticillata*: Past and present distribution in Japan. *Science* 142:1680–81.

PART IV

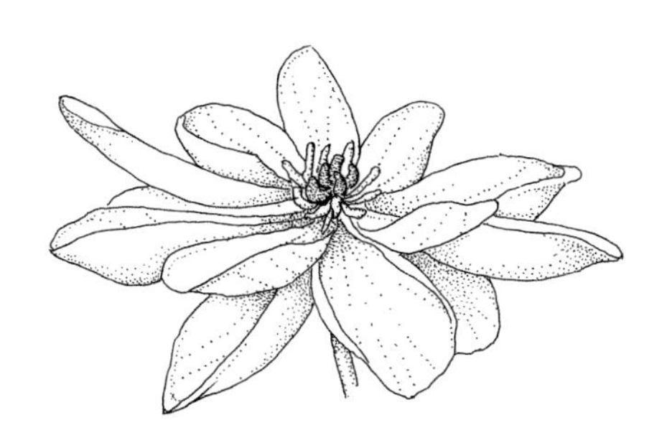

CHAPTER 14

Coming into Flower

A revolution took place in the plant kingdom in the mid-Cretaceous that forever changed the course of evolution and thus life on earth: the old guard of the gymnosperms was overthrown by the newcomer angiosperms. Within 133 million years the angiosperms, also known as the flowering plants, overpowered all other plant groups and reigned over the global flora. They are the most diverse, abundant, and ecologically variable plant group. Angiosperms represent up to 90 percent of all higher plants and are indispensable to our everyday lives.

Without angiosperms, most of the food plants that we subsist on today would not exist, because most of our cultivated species are flowering plants. The nonhuman animal world would collapse too. Cows and sheep would not be feeding on green meadows, zebras would not be browsing on savanna grass, and giraffes would not be nibbling on acacia leaves. The global insect fauna would be completely different as well if the insects that pollinate flowers today no longer had their partner flowers. In fact, the world's biomes are so strongly defined by flowering plants that it would be difficult to imagine a time without grasslands, deciduous hardwood forests, tropical forests, or desert cacti (fig. 14.1).

Yet, before the Lower Cretaceous there were no angiosperms. Instead, the world comprised the lush greens and browns of ferns, conifers, cycads, ginkgoes, and other gymnosperms. It is generally believed that one of the Mesozoic gymnosperms, perhaps a seed fern of some sort, gave rise to the flowering plants. However, no one really knows where they came from, despite intense research in paleobotany. Almost a century and a half ago, this question even perplexed the great biologist and evolutionist Charles Darwin. In 1879 Darwin wrote to his friend and botanist colleague Joseph Hooker exclaiming that the origin of the flowering plants was "an abominable mystery." Darwin was deeply puzzled because it seemed that the flowering plants appeared fully formed in the fossil record quite suddenly, without any transitional forms or evidence that they had descended from other major plant groups. To this day their origins remain a mystery.

The earliest evidence of the flowering plants in the fossil record is 133-million-year-old pollen from the Lower Cretaceous of Israel. Angiosperm pollen is structurally quite unlike gymnosperm pollen in having a different wall structure and surface features. In fact, Darwin's 1879 observation remains acutely relevant today. Angiosperm pollen is so different that it seem to have suddenly arisen from nowhere.

It is thought that the first angiosperms may have been weedy pioneers that colonized unstable environments in the open light. Such environments would include, for example, point bars or riverbanks where the vegetation is regularly washed away or drowned by flooding. This would explain why evidence of the oldest flowering plants is so hard to find in the fossil record. Point bars and riverbanks usually get eroded away and do not usually preserve fossils. A new idea offers a twist to this hypothesis. It has recently been suggested that the earliest angiosperms inhabited dimly lit, disturbed forest understories or shady streamside settings, where the chances of preservation would also have been low.

Figure 14.1. The unmistakable form of a giant saguaro cactus, an angiosperm, at Saguaro National Park in Arizona. Photo from author's archives.

In regard to fossil flowers, the first ones showed up about 125–30 million years ago, a few million years after angiosperm pollen appeared on the scene. These flowers were tiny, measuring only a few millimeters (about a tenth of an inch) (see fig. 15.6). In fact, the entire plant of the earliest angiosperms, including the stem and leaves, was small and not woody.

Ironically, some of the earliest fossil flowers discovered are exquisite in their preservation due to wildfire. The charring of the floral structures at low temperatures before they entered the sediment record preserved them in three dimensions with complete cellular fossilization. Adding to their good luck was that the flowers were deposited in loose sand grains that were never glued together to form solid rock. Thus, the tiny, delicate fossils were never crushed or flattened. The minute fossil flowers, as well as fossil fruits, could then be carefully sieved out of the sand in perfect form. Small charred fossil flowers that are 125 million years old occur at only two sites in the world: in Skåne (Scania) in southern Sweden and near Torres Vedras in Portugal. Recently, even older fossil flowers of a fully aquatic angiosperm have been reported from slightly older, 130-million-year-old limestones in the Pyrenees of Spain.

What Makes an Angiosperm?

Flowering plants are identified by a suite of six characteristics. The most obvious feature is that they bear flowers. Most flowers have four sets of flower parts: green sepals, colored petals, pollen-bearing structures called stamens, and the seed-containing structures known as carpels. Not all flowers are colorful; wind-pollinated flowers are usually small, inconspicuous, and green in color and have fewer or even no petals and sepals. However, many flowers flaunt their glorious floral colors and flashy shapes to signal and attract pollinators (fig. 14.2). Large sizes, bold hues, and striking morphologies in flowers have also been enthusiastically selected for by humans.

Figure 14.2. Flowers come in a dazzling variety of colors and morphologies. *Left, Ophrys lesbis*, the Lesvos orchid; *top center, Ipomoea tricolor* "Heavenly Blue," a blue-colored morning glory; *top right, Hypericum* "Hidcote," St. John's wort; *bottom center*, a pink-flowering *Nymphaea* cultivar; *bottom right, Caesalpinia pulcherrima*, the dwarf poinciana. Photos by author.

The carpel, the central reproductive organ that encloses the seeds, is a feature only found in angiosperms. The combining form *angio-* is derived from *angeîon*, the ancient Greek word for "vessel," while *sperm* comes from *spermos*, meaning "seed." In angiosperms, the seed is enclosed in a (protective) vessel, the carpel, while the gymnosperms have "naked seeds" that are commonly borne in a cone (see chapter 7).

The next set of distinctive features are found in angiosperm leaves, wood, and pollen. Angiosperm leaves often have a fine, net-like venation, while the wood has large-diameter water-conducting cells that enable the tree to pull incredible amounts of water very quickly from its roots to its leaves. Angiosperm pollen is so different from gymnosperm pollen in its wall structure and surface features that specialists can identify it at a glance using a light microscope.

Lastly, a fertilization process that is unique in the plant kingdom connects all flowering plants. In this process, called double fertilization, there is a twofold, simultaneous fertilization of the egg nucleus and the endosperm nuclei. Each pollen grain that outcompetes the other pollen grains to produce a seed contains two sperm cells. The pollen grain lands on the flower's stigma and grows a pollen tube as quickly as possible to reach the ovule, or egg cell. If the pollen grain tube wins the race and reaches the ovule first, one of its sperm cells fertilizes the egg nucleus in the ovule, which eventually becomes the seed. The second sperm cell then fuses with the two so-called polar nuclei to form the endosperm, a special tissue in the seed that will nourish the embryo plant. This complex process ensures that there is no waste of energy; the endosperm is only produced if the ovule is fertilized. This reproductive strategy is different in cycads, in which the endosperm is produced no matter what, throwing away the energy the cycad could have invested instead in green growth or another attempt at reproduction. Unfortunately for paleobotanists, double fertilization is the one feature of flowering plants that has not yet been identified in the fossil record.

The Appearance of the First Flowers

A recent study using a massive new data set and a state-of-the-art analytical system modeled what an ancestral angiosperm flower would have looked like (illustrated at the beginning of this chapter). This study reconstructed the first flower as radially symmetrical and having both sexes contained in each flower. The sepals and petals were undifferentiated from one another, known collectively as tepals, and resembled ordinary petals as we think of them today. The tepals and the next, inner set of flower organs, the pollen-bearing stamens, were organized into two or more concentric circles. Only the separate carpels in the center of the flower were spirally arranged. In many ways, this ancestral flower may have resembled a typical water lily flower of *Nymphaea* (chapter 15). However, for botanists the interesting evolutionary insight is provided by the details. Today, *Nymphaea* flowers differ in having sepals that may gradually take the shape of petals (not uniform-appearing tepals), flower parts arranged in spirals (not whorls), and fused carpels in the center of the flower (not separate carpels). These differences between the hypothetical ancestral flower and the geologically oldest water lily flower imply that the angiosperms may have appeared and evolved morphologically long before fossil flowers showed up in the rock record 130 million years ago.

Coming into Flower in the Cretaceous

After the first flowers appeared in the Lower Cretaceous, angiosperms underwent an intense period of diversification during the Upper Cretaceous in which they sought to fill all ecological niches. Angiosperms such as the water lilies and lotuses returned to the water, where they sought dominance of the freshwater realm (chapters 15 and 16). A few, such as the mangrove palm, *Nypa* (chapter 17), returned to the coastal marine realm, where they mastered the harsh conditions of the intertidal zone. Seagrasses went even further into the ocean, spreading globally into the fully marine realm, and thus became essential as the base of the food chain for herbivorous marine life such as green turtles, dugongs, manatees, fish, sea urchins, and crabs.

Most flowering plants went in the other direction, though, staying terrestrial but becoming larger and woody. From their small and inconspicuous starting sizes on point bars or the dimly lit forest floor, they grew in stature to become woody shrubs, small trees, and larger trees. Throughout the Upper Cretaceous and the Cenozoic, tall and robust angiosperm trees challenged the supremacy of the conifers in a fight for tropical, subtropical, and warm temperate zones. In the end, they managed to restrict the natural distribution of conifer forests to the boreal regions.

Some flowering plants stayed herbaceous. These nonwoody forms were content growing in the dark forest understory or moved into open or drier environments such as meadows, savannas, or deserts. Meanwhile, angiosperms as a whole enticed insects, birds, and even mammals, such as bats, to become their partners in pollination. Flowers offered pollinators nectar, pollen, and even perfume and oils, while the plants benefitted from improvements to their genetic fitness through the exchange of genes with other individuals of the same species. Through the coevolution of angiosperms and insects, primarily bees, more than 87 percent of all angiosperms rely on animals for their breeding program. And of course the animal world depends on flowering plants for more than just nectar or pollen. Humans, for example, eat angiosperm flowers, fruits, leaves, stems, and even roots.

It is hard to think of such a clever and successful group of plants, one that diversified so quickly that all major groups of angiosperms appeared before the end of the Cretaceous, 66 million years ago, as survivors of paleontological time. Yet, some families emerged early on in the first stages of angiosperm evolution, and distinctive genera have survived intact until today. The genera *Nymphaea, Nuphar*, *Nelumbo*, and *Nypa* (chapters 15–17) have maintained their morphological identity through time. They have found creative new ways to lure pollinators or disperse their fruits. These plants have also successfully fought to thrive in their chosen habitats since the Upper Cretaceous. For these reasons, they too are tenacious plant survivors from deep time.

FURTHER READING

BEEINFOrmed_7_The_Importance-of-Insect-Pollinatorssjlouz8ql.pdf. 2018. https://beecare.bayer.com/media-center/publications.

Field, T. S., N. C. Arens, J. A. Doyle, T. E. Dawson, and M. J. Donoghue. 2004. Dark and disturbed: A new image of early angiosperm ecology. *Paleobiology* 30:82–107.

Friis, E. M., P. R. Crane, and K. R. Pedersen. 2011. *Early Flowers and Angiosperm Evolution*. Cambridge: Cambridge University Press.

Gomez, B., V. Daviero-Gomeza, C. Coiffard, C. Martín-Closas, and D. L. Dilcher. 2015. *Montsechia*, an ancient aquatic angiosperm. *Proceedings of the National Academy of Sciences* 112:10985–88.

Sauquet, H., M. von Balthazar, S. Magallón, J. A. Doyle, P. K. Endress, E. J. Bailes, E. Barroso de Morais, K. Bull-Hereñu, L. Carrive, M. Chartier, G. Chomicki, M. Coiro, R. Cornette, J. H. L. El Ottra, C. Epicoco, C. S. P. Foster, F. Jabbour, A. Haevermans, T. Haevermans, R. Hernández, S. A. Little, S. Löfstrand, J. A. Luna, J. Massoni, S. Nadot, S. Pamper, C. Prieu, E. Reyes, P. dos Santos, K. M. Schoonderwoerd, S. Sontag, A. Soulebeau, Y. Staedler, G. F. Tschan, A. W.-S. Leung, and J. Schönenberger. 2017. The ancestral flower of angiosperms and its early diversification. *Nature Communications* 8:16047.

Taylor, D. W., and L. J. Hickey. 1996. Evidence for and implications of an herbaceous origin for angiosperms. In *Flowering Plant Origin, Evolution & Phylogeny*, ed. Taylor and Hickey, 232–66. New York: Chapman & Hall.

Nymphaea odorata

CHAPTER 15

The Water Lilies *Nymphaea* and *Nuphar*

Is there a flower that evokes the joy of a summer's day better than a water lily? Bold, cheerful, photogenic, abundant, and sweetly fragrant, water lily flowers blossom all summer long. Yet, they are a bit enigmatic, floating serenely as they do on the water, when nearly all other summer flowers are firmly anchored in the ground. Water lily flowers are saucer-shaped, with many spiky petals on the outside and numerous saffron yellow stamens in the center (fig. 15.1). They are brightly colored, ranging from stark white to soft yellow, fuchsia pink to bright red and orange, deep blue to violet. The flowers of some water lilies, called "changeable," are a different color each day. They can start out as a light yellow on the first day, deepen to a reddish-orange on following days, and finally reach a dark red.

Water lilies are popular with gardeners, horticulturists, botanists, nonbotanists, and even painters. Their charisma lies in their stunningly beautiful flowers, which open continuously for months (fig. 15.2). French artist Claude Monet was so fascinated with water lilies that he spent thirty years depicting the same scene in 250 paintings,

Figure 15.1. The blue, or star, lotus, *Nymphaea nouchali*, which is native to Asia, growing at the Shangri La Botanical Garden in Orange, Texas. Just above the opened flower is a flower bud, and clinging to the lower petals of the opened flower is a dragonfly nymph. Photo courtesy of Cynthia Parish Photography.

over and over again: dreamy impressions of water lilies floating on his ponds in Giverny, France. The mood in his painting shifts with the colors and shadows reflected on the water's surface and the interplay of light and shade. The water lily motifs are commonly framed by his Japanese bridge or by weeping boughs of willow leaves. However, what is constant is the bright, sharp-petaled water lilies and their somber-colored lily pads.

Water lily flowers and their leaves are buoyant and usually float on the water's surface, although some flowers may be held slightly above the water's surface by a stiff, slender flower stalk (fig. 15.3). Water lily plants prefer quiet freshwater ponds and lakes with little wind or water disturbance. They do best in ponds that get several hours of full sun every day, and many species like warm water temperatures. Water lilies can tolerate very little salt in the water, so they are good indicators of ecological conditions such as freshwater, and not brackish water. All water lilies are perennials. In harsher climes, the leaves die off as winter

Figure 15.2. Water lily flowers and lily pads of *Nymphaea* "William Falconer," which blossom all summer while floating on the water at the Bonn University Botanic Gardens, Germany. Photo by author.

Figure 15.3. *Nymphaea odorata*, the American white water lily, in which some flowers are held upright above the water's surface, growing here in the Sabine River, Louisiana. Photo courtesy of Cynthia Parish Photography.

approaches, and the horizontal stems at the bottom of the body of water generate new leaves the following spring.

Water lilies have a global distribution, although there are morphological and growth differences between the temperate and tropical species. One difference, besides the temperate plants being winter hardy, lies in their leaves. Temperate water lily leaves are round or oval and have a deep cleft at their base. In elongate leaves, sometimes the basal notch and the rounded basal lobes produce a heart shape. By contrast, the leaves of tropical water lilies may have a toothed or scalloped margin that turns up at the rim, and they are commonly speckled or mottled with red. The leaves of tropical species can also grow huge, sometimes reaching more than 3 m (nearly 10 ft) in diameter. Recently, the discovery of a new species of the famous tropical water lily genus *Victoria*, the one that is commonly photographed with a baby sitting on its giant floating leaf, has now brought the number of species in the genus to three. As in the discovery stories of *Wollemia* (chapter 10) and *Metasequoia* (chapter 12), a persistent, sharp-eyed botanist recognized the subtle distinctions of the new species. In this new age of molecular biology, however, it was genomic sequencing that confirmed that this plant was truly a new type of water lily. The species was then formally described and named *Victoria boliviana*, honoring its country of origin in the New World tropics.

Water lily leaves primarily serve the plant by photosynthesizing, but they are beneficial for pond life as well. One way in which water lily leaves, also called by the sweet fairy-tale name "lily pads," help pond animals is by offering shelter (think of frogs on lily pads) and protection (fish hiding from piscivorous birds). Another way the water lily leaves support the ecological health of a pond is to shade the surface of the water, which slows the growth of algal pond scum. Water lily plants are anchored by their rhizomatous stems growing in the mud on the bottom of ponds and lakes. Their flowers and leaves grow upward through the water, tethered to the stem by a long, flexible stalk.

The water lily family Nymphaeaceae comprises five genera. The best known, especially in the temperate zones, are the white water lily, *Nymphaea*, and the yellow water lily, *Nuphar*. The sacred lotus, *Nelumbo* (chapter 16), was once considered a member of this family, but genetic studies show that the sacred lotus is actually more closely related to proteas and sycamore trees.

There are two common species of white-flowering water lilies, one in the New World and the other in the Old World. The New World species, *Nymphaea odorata* (see fig. 15.3), is widespread in North America, with a huge range from northern Canada to Central America. This species has also been reported from Brazil and Guyana. As its species name, *odorata*, suggests, it is a very sweetly scented flower; it is sometimes called the fragrant waterlily. The Old World, white-flowering species *Nymphaea alba* is also very widespread in its biogeographic range. This species is found all over Europe, from Scandinavia to the Mediterranean, in the Middle East and Africa, and in tropical and temperate Asia. *Nymphaea alba* was the first species of the genus *Nymphaea* described and published by Carl Linnaeus in his *Species Plantarum* in 1753.

While white-flowering water lilies are widespread, the avid cultivation and breeding of *Nymphaea* since the late 1800s have produced flowers of all hues. Water lily hybridization not only created new colors in the flowers and leaves (fig. 15.4) but also increased the number of petals and stamen to make the individual blooms showier.

Despite its sweet scent and flashy flowers to attract pollinators, *Nymphaea odorata* has a devious way to ensure that cross pollination happens in every flower. While *Nymphaea* flowers are not pollinator-specific—indeed, a great variety of insects visit water lilies, from flies to beetles to various types of bees—seed set can only occur if pollen is deposited by a visiting insect.

On the first day a flower of *Nymphaea odorata* opens, only its female parts are mature. The saucer shape of the flower is perfect for holding a bowl of viscous liquid in its center. Any insect that falls into the liquid will sink to the bottom of this bowl, and any pollen on the insect's body will be washed off into the liquid. The pollen, presumably from a different water lily flower, then drifts down through the liquid to the receptive female parts to achieve cross-pollination. Having washed off the old pollen, the insect crawls out of the liquid unscathed and then out of the fully opened flower. The insect is now dusted with fresh pollen to take to the next flower. After that first day, the flower closes in the evening and reopens the next day, with the female parts no longer receptive and, instead, the male parts yielding pollen in a mature state.

Figure 15.4. A *Nymphaea* cultivar with pale petals that deepen into a hot pink color at their tips and dramatically dark red, spotted leaves growing at the Missouri Botanical Garden in St. Louis. Photo by author.

After a water lily flower has been successfully fertilized, the flower stalk begins to contract like a spring. This action draws the developing fruit underwater. This ensures that the seeds can develop safely, away from plant predators. Water lily seeds are nutritious, as they are high in starch, protein, and oil. These seeds, as well as water lily flowers and rhizomes, were a major food source for the native people of Australia, for example.

The yellow water lily, *Nuphar*, is known by the buttercup-yellow color of its flowers. The best-known species is *Nuphar lutea*, which is also called the yellow pond lily, the spatterdock, or even the brandy bottle. This last name comes from the stale smell of alcohol that is emitted by the flowers. Flies are particularly attracted to this scent and are the most common pollinators, although beetles and bees also visit *Nuphar* flowers. Unlike *Nymphaea* flowers, with petals in a wide array

Figure 15.5. *Left*, the least water lily, also called the small yellow water lily, *Nuphar pumila*, at the Bonn University Botanic Gardens in Germany. The largest flowers are 3 cm (just over 1 in) across, while the leaves look like little green floating hearts. *Right*, a freshy picked fruit of *Nuphar lutea*, the yellow water lily, which resembles a miniature green bottle in a shallow cup. The flower cup measures 5 cm (ca. 2 in). Photos by author.

of colors, all species of *Nuphar* have yellow blossoms. *Nuphar* flowers also differ from those of *Nymphaea* in having shorter petals with rounded tips. The flowers of *Nuphar* look more like a little yellow cup than a saucer. The leaves of the yellow water lily are similar to those of the common water lily in having a notched base and floating on the water's surface. However, the leaves of *Nuphar* are more elongate and have a sharper tip, giving them more of a cordate, or heart-shaped, appearance (fig. 15.5, left).

Like the flowers of *Nymphaea*, those of *Nuphar* must be pollinated by insects in order to set seed. To avoid self-pollination and to ensure cross-pollination, the female parts in a *Nuphar* flower develop first, a day before the male parts with pollen. However, *Nuphar* does not employ a water trap to wash pollen off of unsuspecting insects. Instead, it uses other methods to get the insects to do its bidding.

An insect is first lured by the sweet aroma of a nectar reward. To reach the nectar, the insect must enter a small triangular opening in a closed *Nuphar* flower and crawl across a mucous-covered stigmatic disk, depositing any pollen it may have collected from a different plant. On the second day, the yellow *Nuphar* flower will unfurl its petals into a little cup shape, releasing the insect and offering it pollen as a consolation prize, as its first row of anthers is now mature and offers ripe pollen. The mucus on the stigma dries up, signaling the lack of receptiveness of the female parts. The pollen-bearing parts continue to mature, row by row, until all rows have presented pollen, and fruit development begins. Unlike with *Nymphaea*, which hides its developing fruits and seeds underwater, *Nuphar* fruits stay on the water's surface. *Nuphar* fruits look like a green, round-bottomed flask nestled in a little green cup (fig. 15.5, right).

The popularity of lilies such as *Nymphaea* and *Nuphar* with plant scientists goes beyond their admiration of the plants' aesthetic beauty; it turns out that water lilies are among the oldest flowering plants on earth. While they may not be as old as horsetails, ferns, or cycads, water lilies first appeared at the very advent of the angiosperms in the Lower Cretaceous, about 115 million years ago. One of the earliest occurrences of fossil water lily flowers and pollen is known from the Aptian–Albian of Portugal, while leaves have been found in a sedimentary sequence of similar geological age in the eastern United States.

The fossil water lilies in Portugal consist mainly of the central part of the flower. The seed-containing gynoecium is preserved, along with the stigmatic disk on top. Along the sides of the gynoecium are the attachment scars of the stamens, petals, and sepals. These scars make it evident that this flower had whorls of many flower parts, just as flowers of modern water lilies do. The most interesting thing is that these fossil water lily flowers are really tiny, only 3 mm (one tenth of an inch) high.

Figure 15.6. A single flower of *Nymphaea* "William Falconer," nearly 13 cm (ca. 5 in) in diameter, at the Bonn University Botanic Gardens in Germany. *Inset*: for comparison, the size the first water lily flower in the Lower Cretaceous of Portugal would have been. Photo by author.

The Cretaceous water lily flower would be absolutely dwarfed by a normal-sized *Nymphaea* flower (fig. 15.6).

Like the fossil flowers, water lily leaves from the Lower Cretaceous of eastern North America are also readily identifiable by their morphology. The leaves are round and have a notch at their base. Their venation is also similar to that of living leaves, with veins that radiate from the base. The oldest water lily leaves in these geological sediments generally have poorly organized venation. However, as time and evolution go on, the venation in the fossil leaves preserved in younger sediments becomes more regular and increasingly more nymphaeaceaous. In rocks of the same age in Brazil, an entire water lily plant was found with rhizomes and young emerging leaves. Described as a member of the family Nymphaeaceae, this plant's leaves are heart-shaped with a shallow notch. They also display the basal radiating veins so characteristic of today's water lily leaves.

The water lily fossils from the same geological time slice from Europe, North America, and South America show that the water lily family moved quickly to colonize bodies of freshwater in different parts of the world. These plants appeared only 119 million years after the first appearance of angiosperm pollen. The water lilies took immediate advantage of the open ecological niche, moving into the freshwater realm early on, while most other early angiosperms remained firmly rooted on land. These fossil water lilies differed enough from their living relatives that they could not be assigned to a modern genus. Yet, they show that the order Nymphaeales and the family Nymphaeaceae can be traced back deep into the Lower Cretaceous.

The lineages of the modern genera *Nymphaea* and *Nuphar* emerged not too long afterwards. *Nymphaea* leaves appeared in the Upper Cretaceous, about 90 million years ago. By contrast, *Nuphar* seeds have been found from the beginning of the Cenozoic, about 62 million years ago. After that, fossil *Nymphaea* leaves occurred in freshwater lake sediments in the Cenozoic throughout western North America and Europe. One particularly lovely water lily leaf from the Oligocene of France is called *Nymphaea calophylla* (fig. 15.7). Its species name is derived from the prefix *calo-*, from the Greek *kalos*, meaning "beautiful," and the Greek suffix *phyllus*, meaning "leaved."

As will be discussed in chapter 17, on the mangrove palm, *Nypa*, the temperatures of the Cenozoic reached their high point during the

Figure 15.7. A fossil leaf of *Nymphaea calophylla* from the Oligocene of southern France, framed and hung high as an object of wonder and beauty on a wall of the Museum d'Histoire Naturelle in Aix-en-Provence, France. The leaf is remarkably large, measuring 30.6 cm (12 in) across. Photo by author.

Eocene. A very broad belt of tropical conditions—warm, humid, and equable—extended far north and far south of the equator. These optimal growth conditions must have encouraged the speciation of new lineages of water lilies, as well as the geographic expansion of *Nymphaea* and *Nuphar*.

In the Eocene of Washington State, a large assemblage of plant parts pertaining to the modern genus *Nuphar* was found that represents the most complete plant species of the yellow water lily in the fossil record. The fossil remains consist of several fruits, one of which bears seeds, as well as stamens and stamen attachment scars. Dispersed seeds, a stigmatic disk, petals and sepals, and rhizomes were also found nearby. Only the leaves were missing.

Today, one interesting difference between living *Nuphar* and *Nymphaea* is their biogeographic distribution. While *Nymphaea* has a global distribution, *Nuphar* is restricted to the Northern Hemisphere. This makes it all the more striking that a close relative of *Nuphar* once grew in Antarctica some 40 million years ago. Fossil seeds similar to those of living *Nuphar* were found in Eocene sediments on one of the islands that surround the Antarctic Peninsula. However, because some cellular features were intermediate between the yellow water lily and the closely related aquatic genera of the watershield family, Cabombaceae, the fossil seeds were described as being from an extinct and intermediary water lily genus, *Notonuphar*.

Another interesting type of fossil water lily was recently described from the Eocene of Germany. In ancient Lake Messel, near Frankfurt, 48-million-year-old water lily leaves were discovered that are similar in their elongate shape to those of *Nuphar*, but with venation that is closer to that of *Nymphaea* (fig. 15.8). On the basis of an extensive, computer-supported analysis of 55 leaf characteristics found in 34 species of water lily leaves, these fossil leaves were also recognized as belonging to a new, extinct, intermediary genus of water lilies. The fossil leaves were named *Nuphaea* to show its transitional relationship to living *Nuphar* and *Nymphaea*. This new Eocene genus offers another example of the greater biodiversity of water lilies in the paleontological past.

Water lilies have a long and illustrious fossil history extending back to the dawn of the angiosperms. The lineages of *Nymphaea* and *Nuphar* emerged as early as the Upper Cretaceous, and the biodiversity of the water lilies increased in the Cenozoic. During the peak of global tropical conditions in the Eocene, 40 million years ago, new types of

Figure 15.8. Reconstruction of the open-water habitat in Lake Messel, near Frankfurt, Germany, during the Eocene, with *Nuphaea* water lilies, 48 million years ago. Photo courtesy of HLMD (Hessisches Landesmuseum Darmstadt), Oscar Sanisidro.

water lilies evolved, although not all of these lineages made it to the present day. The white water lily, *Nymphaea*, and the yellow water lily, *Nuphar*, however, are two tenacious freshwater flowering plants that did survive the hundreds of millions of years until now. Whether their success came from exploiting the freshwater ecological niche, enticing and trapping pollinators to do their bidding, or both, water lilies are among the most widespread, popular, and successful flowering plants in the aquatic realm today.

FURTHER READING

DeVore, M. L., W. Taylor, and K. Pigg. 2015. *Nuphar carlquistii* sp. nov. (Nymphaeaceae): A water lily from the latest early Eocene, Republic, Washington. *International Journal of Plant Sciences* 176:365–77.

Friis, E. M., A. Iglesias, M. A. Reguero, and T. Mörs. 2017. *Notonuphar antarctica*, an extinct water lily (Nymphaeales) from the Eocene of Antarctica. *Plant Systematics and Evolution* 303:969–80.

Friis, E. M., K. R. Pedersen, and P. R. Crane. 2001. Fossil evidence of water lilies (Nymphaeales) in the Early Cretaceous. *Nature* 410, no. 6826: 357–60.

Gee, C. T., and D. W. Taylor. 2019. An extinct transitional leaf genus of Nymphaeaceae from the Eocene lake at Messel, Germany: *Nuphaea engelhardtii* Gee et David W. Taylor gen. et sp. nov. *International Journal of Plant Sciences* 180:724–36.

Horton, H. 2022. Newly identified waterlily species is world's largest. *Guardian*, 4 July 2022. https://www.theguardian.com/environment/2022/jul/04/third-species-of-giant-waterlily-discovered-at-kew-gardens.

Nelumbo nucifera

CHAPTER 16

The Sacred Lotus, *Nelumbo*

One of the most beloved plants in Asia is the sacred lotus. A profusely flowering, freshwater plant with beautiful pink and white blossoms, large leaves, tasty seeds, and nutritious roots, the sacred lotus is as important in religion, art, and culture as it is in culinary and economic ways. One of the oldest symbols on earth, the sacred lotus is a symbol of purity. This is because lotus leaves and flowers always rise up fresh and clean (fig. 16.1), unsullied by the mud of the dirtiest ponds. The sacred lotus is also a symbol of longevity, for lotus seeds can still germinate after a thousand years.

The leaves, flowers, flower buds, and even fruits of the lotus are large, showy, and attractive. In particular, lotus flowers and leaves are a popular design motif. These appear in paintings and fabrics, as well as on three-dimensional objects like candles, vases, drinking vessels, and incense holders (fig. 16.2). Because of the open nature of the mature flower and the exquisite form of its spreading petals, the lotus flower is often used in statues as the base of a deity (fig. 16.3). In the case of a bodhisattva emerging from the flower, the lotus also symbolizes enlightenment, self-regeneration, and rebirth.

Figure 16.1. The sacred lotus, *Nelumbo nucifera,* growing in the Huntington Botanical Gardens in San Marino, California. *Left,* a delicately colored flower on a stalk rising high above the aerial leaves. The neighboring flower bud and young fruit show the changes in size and form that occur during the development of the reproductive structures. *Right,* the interior of an *N. nucifera* flower, with the bright yellow, developing fruit surrounded by a ring of stamens. Photos by author.

Nearly all the parts of a lotus plant can be eaten or used in food preparation. This includes the seeds, the leaves, and the roots. For the Mid-Autumn Festival in China, the seeds are boiled, puréed, and lightly sweetened to be made into one of the traditional fillings of the popular mooncakes. Lotus seeds give the pastries a slightly nutty taste. The protein-rich seeds can also be prepared like popcorn and enjoyed whole. The large, disk-shaped leaves of the lotus plant, on the other hand, are sometimes employed to hold packets of steamed sticky rice eaten as dim sum.

Figure 16.2. (*Above*) The sacred lotus as a sculptural motif. *Left*, an ancient jade cup in the shape of an upturned lotus leaf, on display in the National Palace Museum in Taipei, Taiwan. *Right*, incense holder adorned with frogs and various motifs of the sacred lotus: on the cap, a fully opened flower with developing fruit atop the leaves; on the handles, young fruits; along the side, flowers and pond frogs. Photos by author.

Figure 16.3. (*Left*) A bodhisattva holding a child while seated on a stylized lotus flower. From the Ming Dynasty, on display in the National Palace Museum in Taipei, Taiwan. Photo by author.

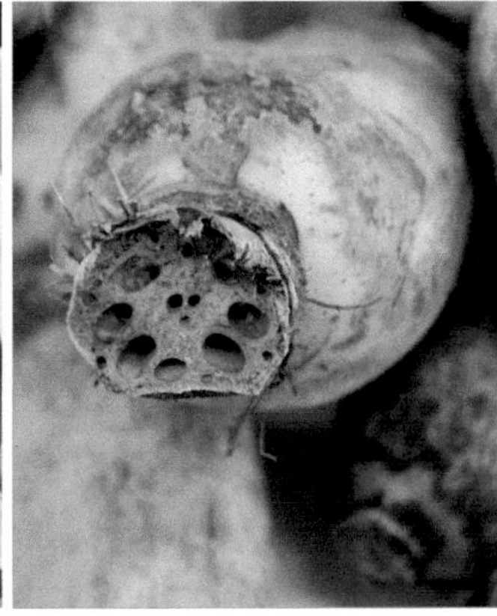

Figure 16.4. *Left*, lotus roots for sale at a market stand in Hong Kong. *Right*, the end of what is technically not the root but the rhizome or stem of a lotus plant, showing its variously sized air passageways. Photos by author.

The part of the sacred lotus most commonly used in cooking is the so-called lotus root. While it would be technically designated as a rhizome, a tuber, or a horizontal stem, lotus root can be readily identified when sliced by its many holes of varying sizes (fig. 16.4). The holes serve as passageways for the circulation of oxygen in the rhizomes, which generally grow underwater in dense mud. The lotus root is characterized by a crunchy texture and can be prepared by boiling, baking, roasting, or stir-frying. It also shows up as a main ingredient in soups (fig. 16.5), as well as soaked in syrup, pickled in vinegar and spices, or even as tea.

The sacred lotus is frequently mistaken for a water lily. While indeed an aquatic macrophyte with bright flowers and round leaves, the lotus is not a member of the order Nymphaeales or the family Nymphaeaceae (see chapter 15). Instead, it occupies its own family, the Nelumbonaceae, in the order Proteales. Thus, the sacred lotus in the family Nelumbonaceae is more closely related to proteas in the

Proteaceae and even to sycamores in the Platanaceae than to true water lilies.

Today, there is only one genus, *Nelumbo*, which is represented by two species. *Nelumbo nucifera* is the sacred lotus known from Asia, while *Nelumbo lutea* is the yellow lotus from North and Central America. The most conspicuous difference between the two species is the color of their flowers. The blossoms of the sacred lotus are white, pink, or gradations in-between, whereas those of the American species are a creamy yellow. However, as it is a popular pond plant, more than 600 cultivars of lotus bearing variously colored flowers have been developed. For example, a cross between the sacred lotus and the yellow lotus species has produced a hybrid with soft, pastel yellow petals with a pink blush.

Figure 16.5. Lotus root soup, a common dish in China, offered for sale here at a food market in Hubei Province. The pink color of the broth and boiled root slices arises from the pigments and other compounds in the starch of the *Nelumbo* rhizome. Photo by author.

Figure 16.6. Fruits of *Nelumbo nucifera*. *Left*, green lotus seed pods for sale along with other fruits at a roadside stand in China. After seed pods are broken open, the green seeds can be harvested. Lotus seeds are eaten raw or cooked, and lotus seed paste is a popular pastry filling in Asia. Photo by author. *Right*, a mature seed pod with widened seed locule openings. Photo by Maarten Heerlien, Creative Commons Attribution-ShareAlike 2.0 Generic license.

The *Nelumbo* flower is not differentiated into green sepals and colored petals as most other flowers are. Instead, a gradation of form and color makes it difficult to distinguish clearly between these flower parts. The fruit that develops in the center of the pollinated flower is at first bright yellow (see fig. 16.1, right), becomes green (fig. 16.6, left), then darkens into a woody brown (fig. 16.6, right). While some may compare its form to a shower head, it is actually a funnel-shaped seed pod containing multiple seeds. As the seeds grow larger, they begin to extend beyond from their respective locules, or seed cavities, protruding slightly from the top surface of the seed pod. The upper end of the seed is marked with a small air hole that provides each individual seed access to oxygen.

As the fruit matures, each seed locule starts to expand, and the locule holes on the surface of the seed pod begin to widen (see fig. 16.6, right). In the end, the locules are so large that the seeds rattle around loosely in them, and the seeds are dispersed by simply falling out of the fruit. These mature, brown-colored seed pods are popular components of flower arrangements and autumnal decorations.

Both the flowers and the leaves of *Nelumbo* are held high above the surface of the water by sturdy stalks. This is another way in which they differ from true water lilies, as most of the flowers and leaves of *Nymphaea* and *Nuphar* float on the water. Only a few species of *Nymphaea* hold their flowers stiffly a short distance above the water's surface (see fig. 15.3). In contrast, the stalked seed pods of *Nelumbo* still remain erect above the water's surface when ripe in the fall. Of course, their shower head–like morphology is also a dead giveaway to the identity of the water plant.

Another difference between lotuses and water lilies resides in the leaves. The leaf stalk, or petiole, of the lotus is attached to the back of the leaf blade at its center. From this central attachment point the leaf veins radiate out toward the leaf's edge, much in the manner of spokes of a wheel. By comparison, the stalk of a water lily leaf is generally attached at the edge of the leaf blade, not at the center.

All leaves, even those of water plants, play an important role in gas exchange. In fact, a good respiratory and circulatory system is crucial for aquatic macrophytes. This is especially true for those with stems and roots immersed in thick, waterlogged mud. In the case of the *Nelumbo*, oxygen enters the plant through its stomata, on the upper side of the leaf lamina. The air is then sent throughout the plant body, from the leaf via the leaf stalk to the rhizome, which is rooted in pond mud. In the rhizome, the oxygen moves through the air passageways that appear as different-sized holes in the cut-open or sliced lotus rhizome. The stream of oxygen then reverses its path, coursing through a different set of air canals from the rhizome back into the petiole and the leaf lamina. From the leaf, the air is expelled through the central plate, a navel-like structure in the center of the leaf on the upper side of the lamina. The central plate is a feature particular to lotus leaves that functions in the ventilation of the plant.

Analogous to the air canals in the rhizomes are special passageways for air circulation in the leaves. These structures are air canals that allow oxygen to circulate on either side of the radiating veins in the center of the leaf. There are even adaptations in *Nelumbo* to contend with damage to its aeration system. Because most of the plant is fully immersed in water, there is always the danger that water may flood the air canals and disrupt the flow of oxygen that is so vital to the well-being of the lotus plant. To minimize any impairment in such a case, the air canals in *Nelumbo* petioles contain peculiar structures that allow air to pass through them but hold back incoming water. The structures resemble irregularly shaped balls with an open, net-like structure and fill the major air canals in the leaf stalk. These so-called petiolar diaphragms are hydrophobic and have been shown to hinder the influx into the air canals of water with a pressure as high as that of a column of water 84 cm (nearly 3 ft) in length.

The leaves of *Nelumbo* are of particular interest to materials scientists because their surface is highly water-repellent (fig. 16.7). The ultrahydrophobicity of lotus leaves comes from extremely small, densely packed protuberances that prevent drops of water, plus any dirt they might contain, from touching and soaking into the leaf surface. Thus, it is the nano-scale architecture of the surface of *Nelumbo* leaves that makes them self-cleaning. This natural phenomenon is known as the lotus effect, and this discovery has been commercially applied to self-cleaning materials and surfaces such as car waxes, paints, roof tiles, and fabrics.

Although there are only two extant species of *Nelumbo*, there are many cultivars, varieties, and breeding lineages. The center of biodiversity of the sacred lotus, *N. nucifera*, is found today in Hubei Province, in central China, where there are many bodies of freshwater near the capital city of Wuhan, such as the Yangtze River and many inland lakes, as well as deep cultural ties to the plant. More than 572 cultivars of *Nelumbo nucifera* have been documented from this region. The cul-

Figure 16.7. Close-up of a leaf of *Nelumbo nucifera* with a spherical bead of water. The leaf surface is so water-repellent that water and other liquids roll off as droplets, leaving the leaf clean and dry. This special property, unique to leaves of *Nelumbo*, is called the "lotus effect." Photo by author.

tivars fall into three categories: rhizome lotus, seed lotus, and flower lotus. The rhizome lotus cultivars were bred to produce a higher yield and a better-quality edible rhizome. The seed lotus cultivars produce fruits with a large number of carpels and have a high seed set. The flower cultivars are generally used for ornamental purposes. They offer a large number of flowers but are the shortest and produce fewer seeds.

These cultivars can produce flowers with a single whorl, one and a half whorls, or two whorls of petals.

Because lotuses have been a staple of East Asian cuisine for several thousand years, many villages in China and Japan have a lotus pond where these plants are cultivated as food. Most lotuses propagate themselves vegetatively through the rigorous growth of their rhizomes. However, lotuses can quickly dominate the water flora of a small pond. To prevent this, the plants can be contained in pots within the body of water to keep the rhizomes from spreading and overwhelming the other aquatic plants. Lotuses also do well in smaller, freestanding tubs. Although *Nelumbo* is winter-hardy, the leaves die back in the fall and reemerge later, in the spring.

Lotus seeds are less commonly an agent of propagation, for their seed coat is extremely thick. The tough seed coat is the reason why lotus seeds found in a dry lake bed in northeastern China dating from more than 1,300 years ago were viable and still capable of germination. To germinate a *Nelumbo* seed, the seed coat must be nicked with a file, sanded off with a grinder, or cut with a fine saw so that water can reach the plant embryo within. When the embryo has absorbed enough water, the seed coat can be removed by hand, and the baby plant allowed to grow.

Because plants are organisms with relatively simple forms, especially their leaves, it is sometimes tricky to identify with certainty a particular leaf in the fossil record. For example, a round leaf with a centrally positioned petiole and radiating venation, such as the leaf of the sacred lotus, is a general leaf type. This type of leaf occurs in several unrelated plant species, including nasturtium, water pennywort, and the Chinese money plant. Thus, it is important to find other distinctive characters in the fossil leaves in order to make a correct identification. In the case of *Nelumbo*, the occurrence of other fossil plant organs, such as lotus-like rhizomes, fruits, and seeds, alongside the leaves strengthens the botanical determination of the fossil foliage.

Numerous examples of round leaves resembling those of *Nelumbo* have been reported from the Cretaceous, starting from about 100 million years ago. Many of these older descriptions, however, did not necessarily take into account features beyond a round leaf shape and radiating veins. Leaves of living lotus, for instance, also have a central plate on the upper side of the leaf, a centrally positioned petiole on the lower side of the leaf, a slightly funnel-shaped lamina, air canals that run parallel on either side of the veins, and veins that loop along, but do not reach, the leaf margin. Upper Cretaceous fossil leaves from North America have been described as having a round shape, radiating veins, a funnel-like form, and looped veins near the leaf edge. However, because the leaves lacked a central disk and differed in other venation details when compared with the leaves of living *Nelumbo*, these fossil leaves were given a new name, *Exnelumbites*. In this case, the recognition of a new genus based on leaf differences documents that there was more diversity in the family Nelumbonaceae 70 million years ago than in the present day. Hence, the much older reports of *Nelumbo* and *Nelumbo*-like fossil leaves so abundant in the Upper Cretaceous of North America and Europe should be examined with fresh, modern eyes to determine whether they pertain to the living genus of *Nelumbo* or to more ancient taxa within the family.

Elsewhere in the world, also roughly 70 million years ago, *Nelumbo* leaves and fruits showed up in Patagonia, Argentina. Unlike the North American leaves, the South American leaves not only were round with radiating veins but had a central plate and looped veins near the margins. Associated with these leaves were fruiting heads resembling the strongly funnel-shaped, shower head–like seed pods found in *Nelumbo* today.

The frequency of lotus leaves in the fossil record increased in the Cenozoic. One striking assemblage of fossil *Nelumbo* leaves, rhizomes, fruits, and seeds was discovered in the Eocene of Hainan Island, China. The hundreds of well-preserved specimens occurring in a single horizon at this locality, along with the presence of adventitious roots attached to the rhizomes, indicate that the fossil plants were unearthed in the ancient pond that they had colonized 40 million years ago.

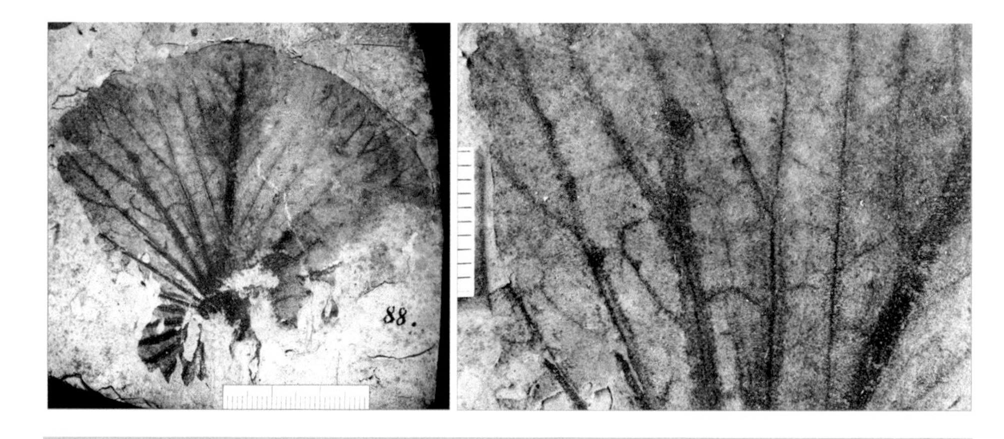

Figure 16.8. A fossil leaf of what may represent a 23-million-year-old lotus from the Oligocene lake at Rott, near Bonn, Germany. *Left*, a fragmentary leaf. Note the similarities to living *Nelumbo* leaves: a round leaf shape, veins radiating from a central position, and a central plate. The entire scale bar measures 3 cm. *Right*, close-up of the venation in the same leaf. Here, the air canals that parallel the veins are visible. The entire scale bar denotes 1 cm (0.4 inch). Photos courtesy of the Museum für Naturkunde, Berlin.

We know with certainty that in the Oligocene lake that existed 23 million years ago near what is today the town of Rott, near Bonn, Germany, true water lilies flourished in the warm water. However, it is becoming clearer that they must have shared this freshwater habitat with lotus plants. In 1855, fossil leaves were described as *Nymphaea lignitica* by naturalists interested in the Rott flora. For nearly 160 years ago it was assumed that these leaves represented true water lilies of the Nymphaeaceae. Recently, a fresh look at the leaves, along with a rigorous phylogenetic analysis of the leaf form and venation, showed that these leaves did not pertain to the Nymphaeaceae. Rather, these fossil leaves belong to the Nelumbonaceae. Like those of living *Nelumbo*, the fossil leaves from Rott are round but lightly funnel-shaped and have a petiole attached mid-lamina, radiating veins, and a central plate. They also show air canals that run on either side of the radiating veins (fig. 16.8).

In summary, like true water lilies, lotuses have a fossil history that extends into the Cretaceous. Fossil leaves assigned to the Nelumbonaceae that differ from the present-day genus of *Nelumbo* show that the biodiversity of the family was greater in the past. Despite global cooling in the Cenozoic, *Nelumbo* has flourished in freshwater lakes over the last 66 million years, which was surely made possible by its special morphological and anatomical adaptations to its low-oxygen habitat.

Today, only two species of the genus *Nelumbo* survive. Nevertheless, the intense, centuries-long domestication of one of the species, *N. nucifera*, the sacred lotus, which has resulted in hundreds of cultivars and a global distribution, coupled with the widespread cultural, culinary, and horticultural popularity, has secured the future of this living fossil.

FURTHER READING

Estrada-Ruiz, E., G. R. Upchurch Jr., J. A. Wolfe, and S. R. S. Cevallos-Ferriz. 2011. Comparative morphology of fossil and extant leaves of Nelumbonaceae, including a new genus from the Late Cretaceous of western North America. *Systematic Botany* 36:337–51.

Gandolfo, M. A., and R. N. Cuneo. 2005. Fossil Nelumbonaceae from the La Colonia Formation (Campanian–Maastrichtian, Upper Cretaceous), Chubut, Patagonia, Argentina. *Review of Palaeobotany and Palynology* 133:169–78.

Li, Y., T. Smith, P. Svetlana, J. Yang, J.-H. Jin, and C.-S. Li. 2014. Paleobiogeography of the lotus plant (Nelumbonaceae: *Nelumbo*) and its bearing on the paleoclimatic changes. *Palaeogeography, Palaeoclimatology, Palaeoecology* 399:284–93.

Shen-Miller, J. 2002. Sacred lotus, the long-living fruits of *China Antique*. *Seed Science Research* 12:131–43.

Vogel, S. 2004. Contributions to the functional anatomy and biology of *Nelumbo nucifera* (Nelumbonaceae). I. Pathways of air circulation. *Plant Systematics and Evolution* 249:9–25.

Nypa fruticans

CHAPTER 17

The Mangrove Palm, *Nypa*

One of the toughest jobs a plant can take on is that of a mangrove: growing along the shores of a tropical estuary. Not only is the climate hot and the substrate salty but the rise and fall of the tides mean a never-ending cycle of freshwater flowing out to the sea coupled with the ebb of marine water rushing landward. Firmly rooted, mangrove plants cannot run or swim away but are forced to tolerate the constant swings in water salinity. Very few plants can tolerate such wide, stressful shifts in salt content, but those that can are actually so well adapted to these harsh conditions that they grow in huge numbers in these environments. In temperate areas, the intertidal zone is dominated by salt marshes, while in the tropics and subtropics it is the mangroves that reign.

Mangroves are tropical trees and shrubs that form dense thickets at the edge of the sea. They are particularly common along coastlines that are protected from rough waves but still inundated daily by ocean waters. These woody plants tend to be anchored in brackish-water muds, but they can start new colonies by invading sandy beaches (fig. 17.1).

Figure 17.1. A typical mangrove plant, growing in tropical North Queensland, Australia. Here the red mangrove, *Rhizophora stylosa*, is colonizing a sandy beach along the Pacific Ocean. The red mangrove has a trunk, prop roots growing downwards from its upper branches, and stilt roots originating from the base of the trunk that loop into the water. Photo by author.

Most people think of mangroves as trees growing in the ocean with an upright trunk and oddly branched, aboveground roots. However, an even stranger plant in the mangrove community is the mangrove palm (fig. 17.2).

The mangrove palm, *Nypa fruticans*, is the only palm tough enough to live in the intertidal environment. For this and other reasons that set it apart, it is the sole member of its own subfamily, the Nypoideae, in the Arecaceae family of palms. Systematically, its isolation is also reflected in the fact that there is also only one genus, *Nypa*, with only one species, *N. fruticans*. Its generic name comes from its local name, nipa, on the Molucca Islands and in the southern Philippines. Contrary to the expectation that it has something to do with reproductive structures, its Latin-derived species name, *fruticans*, refers to its shrubby, thicket-like habit of growth.

Besides thriving under saline conditions, *Nypa fruticans* does not have the usual tall, thin trunk typical of most palm trees. Instead, it has

Figure 17.2. The mangrove palm, *Nypa fruticans*, growing at the edge of a freshwater lake at the Flecker Botanic Gardens in Cairns, Australia. The stem of this palm tree grows horizontally in the mud underwater, so that its fronds appear to spring up magically from the water's surface. Photo by author.

a massive stem that grows horizontally, not vertically, and divides in a regular, forked fashion while buried in the mangrove mud. The long fronds of *Nypa* are attached at their base to this underground stem but then reach up toward the sky, attaining heights of 7 m (23 ft). Partly immersed in brackish water along the fringes of riverbanks, tidal channels, or lagoons, the huge leaves of the mangrove palm seem to shoot straight up from the water's surface. Compare this with the quintessential solitary palm tree on a sandy beach with a long, thin trunk, a crown of fronds, and a dangling bunch of coconuts.

The fruits of *Nypa fruticans* are quite different from the big, round coconuts of the coco palm or the bunches of fleshy dates of the date

Figure 17.3. Two fruiting stalks of the mangrove palm, *Nypa fruticans*, in North Queensland, Australia. On the right, numerous fruits are attached to the stalk to create a large, globose fruiting head. The hard and woody reddish-brown fruits have an angled shape. On the left is the central stalk after the fruits have fallen off into the water and are being dispersed by the currents and tides. Photo by author.

palm. *Nypa* fruits are clustered in large, globose fruiting heads larger than a man's head (fig. 17.3). The flower stalk initially grows upright, but after the flowers are pollinated and fruits begin to form, the entire fruiting head becomes bowed down with the heavy fruits. This head hangs so low that the fruits nearly skim the water's surface. When ripe, the individual fruits fall off, leaving the central stalk with scars where the fruits were once attached.

Each fruit is hard with a thick, woody outer covering, an angular body, and a glossy reddish-brown color. Inside is a fibrous layer that captures air and creates buoyancy. These reproductive parts are dispersed by ocean tides and currents, which transport the floating fruits to new areas to germinate and establish new mangrove palm colonies.

Like the fruits, the pollen of *Nypa* is very distinctive. The pollen grain is made of two halves that fit together into a ball like two clam shells unequal in size (fig. 17.4). The surface of the pollen grain has a granular texture and is covered with spines with broad bases. The suite

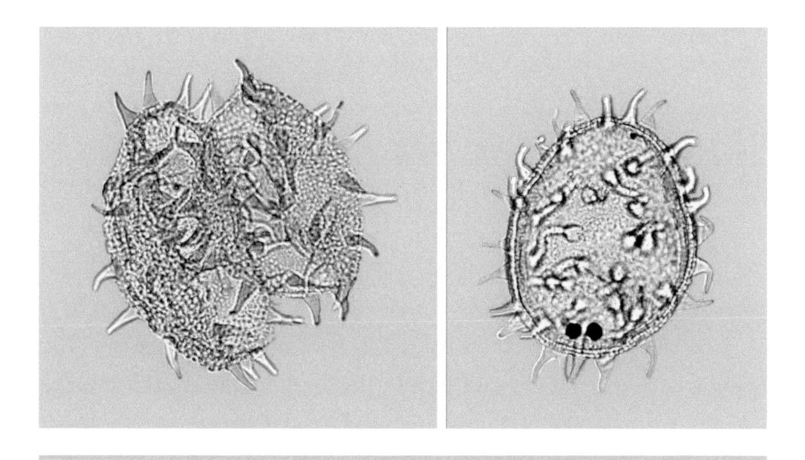

Figure 17.4. Pollen grains of the mangrove palm, *Nypa*, from the middle Eocene of Laredo, Texas. *Left*, a furrow separates the pollen grain into two unequal hemispheres. *Right*, a single hemisphere has become detached from the pollen grain. The pollen is covered by many spines, and the pollen wall has a granular texture. Photos by author.

of unusual characters in the pollen of the mangrove palm makes it easy to identity this type of pollen in the fossil record. In fact, fossil *Nypa* pollen is virtually identical in form and structure to recent pollen, even down to micrometer-sized details of the pollen wall.

The mangrove palm is an Old World species. It is native to a zone that includes both sides of the equator, stretching from southern India in the west, to the Philippines and southern Japan in the northeast, and to southeastern Australia in the southeast. *Nypa fruticans* thrives in hot, humid weather. In its native habitat, the highest temperatures reach, on average, 32–41°C (90–105°F), and the lowest temperatures do not often drop below 20°C (68°F). The monthly rainfall is generally 10 cm (4 in). Thus, mangrove palms like to grow in true tropical climates that are steamy all year round.

Although *Nypa* thrives only within a narrow range of temperatures, it is quite tolerant of salinity. The mangrove palm grows best in brackish water, at an optimum concentration of 1–9 parts per thousand (ppt). Yet, it also thrives in freshwater at 0 ppt and tolerates seawater at the normal salinity of 35 ppt. Its location within the mangrove habitat is mostly determined by other intertidal plants, as it is usually outcompeted by the mangrove species *Rhizophora*, *Avicennia*, *Laguncularia*, and *Sonneratia*. The mangrove palm is commonly rooted in fine-grained, salty substrates that are rich in alluvial silt, clay, and humus. These plants prefer slightly acidic muds with a pH of 5. Like most mangroves, *Nypa* prefers protected areas with little wave action, such as estuaries or shallow lagoons.

Like the plant parts of most palms, the various plant parts of the mangrove palm are widely utilized by the local people. *Nypa* leaves are an important source of thatch for roofs, and they are also used for weaving baskets. Young leaves can be used for wrapping tobacco for smoking.

The sugar-rich sap of young *Nypa* flower stalks is used in a variety of ways. Tapped from the unripe flower stalks below the fruiting head, the sap is commonly the basis of a local alcoholic drink. In the Philippines, the sap is fermented to produce a palm wine that used to be known as *vino de nipa*. Alternatively, it is distilled industrially to make a liquor with a much higher alcohol content called Philippine nipa palm brandy. In Indonesia, the *Nypa* palm sap is fed to pigs in the dry season, which is thought to give the pork meat a sweet flavor.

There is much talk these days about using the sap of the mangrove palm for the large-scale production of biofuels, converting it into ethanol or butanol by fermentation. Mangrove palms would be particularly suited because the sugar-sweet sap they produce in huge quantities is several times more productive as biofuel than sugarcane. Additional benefits include the perennial growth habit of *Nypa*, which allows for year-round harvest of the sap, as well as its growth on nonarable land, where it does not compete with crops. Even the seeds of *Nypa* fruits are ethnobotanically useful. The young endosperm of the seeds is boiled in syrup and eaten as a jelly-like sweet.

Because of the dominance of the true mangrove, *Rhizophora*, today, *Nypa* is sometimes considered to play a minor role in the global mangrove flora. However, in the past *Nypa* had the stage all to itself, as it was the dominant mangrove plant from the Upper Cretaceous to the Eocene. The oldest evidence of *Nypa* pollen comes from the Upper Cretaceous (Campanian), about 75 million years ago, in what is presently Egypt. The emergence of *Nypa* as a palm separate from all other members of the palm family at this time is also supported by genetic data. Ten million years later, in the Maastrichtian, evidence of *Nypa* appeared on a number of landmasses, including Central and South America, Africa, the Indian subcontinent, and parts of Asia. These first appearances occurred at nearly the same time, indicating a rapid, global expansion in biogeographic range by the end of the Mesozoic.

The Upper Cretaceous was a time of intense species diversification and niche expansion for all flowering plants (chapter 14), and the mangrove palm was no exception. *Nypa* was the first angiosperm to return to the sea to colonize the intertidal environment. It is one of the oldest palms and one of the first monocots in the fossil plant record.

While we have seen that some free-sporing plants, such as *Equisetum* (chapter 5) and *Osmundastrum* (chapter 6), and gymnosperms, such as *Ginkgo* (chapter 9), *Araucaria* (chapter 10), and *Metasequoia* (chapter 12), can be recognized on the generic level in deep paleontological time, *Nypa* is one of the oldest fossil angiosperms that can be assigned to a modern genus of the present-day world.

The global heyday of the mangrove palm coincided with the climatic optimum of the Cenozoic in the Eocene, starting 56 million years ago. At that time *Nypa* reached its maximum biogeographic distribution in a broad circumequatorial belt of tropical vegetation. As a result of continental drift, earth's landmasses were arranged differently in Eocene times than they are today. However, the fossil pollen and fruits of *Nypa* could be found as far north as what is currently England, which was then at a paleolatitude of about 45° north, and as far south as the southernmost coastline of what is Tasmania today, at a paleolatitute of about 65° south. Fossil fruits of *Nypa* are commonly found in the Paris Basin (fig. 17.5), a geological region encompassing northeastern France, western Belgium, and southeastern England. This area was a hotspot of marine biodiversity, especially of bivalves and gastropods, during the Eocene.

Contrary to its biogeographic distribution in only the Old World today, *Nypa* was the dominant mangrove in the New World during the Upper Cretaceous and the lower Cenozoic. In Central and South America, mangrove palm fruits and pollen first appeared in the Upper Cretaceous and thrived until they disappeared after the upper Eocene. In North America, *Nypa* fossil fruits and pollen were found in the lower to the upper Eocene. From about 56 million to 36 million years ago, *Nypa* mangroves were widespread from the Gulf Coast, as far west as present-day Texas, through Mississippi, Alabama, and Georgia, to Maryland on the East Coast.

It took some time for paleobotanists to accept the occurrence of mangrove palms and mangroves in the Eocene of Texas. There were several older reports of *Nypa*-like fruits from 1914 to 1931, but the poorly preserved fruits apparently were not convincing. In 1952 a

Figure 17.5. A fossil fruit of *Nypa* from the Eocene of the Paris Basin. Photo by Georg Oleschinski.

beautifully preserved *Nypa* fruit from near Smithville in central Texas was described by a prominent American paleobotanist, Chester Arnold (fig. 17.6), but this too was generally dismissed as a misidentification. Who could imagine a mangrove palm of Indo-Australasia growing 40 million years ago in now landlocked Bastrop County? At this time, no support could be expected from the fossil pollen record, for the Eocene pollen from southern North America was originally misidentified as water lily pollen.

Not until a paleoecological analysis in 1981 did the acceptance of mangrove palm fruits and pollen in North America begin to grow. In that year Norman Frederiksen identified *Nypa* pollen in the Eocene to Oligocene plant communities of the Gulf Coast. These days, we are quite certain about the identification of the mangrove palm in the

Figure 17.6. The top part of a fossil fruit of *Nypa* from the Eocene Weches Formation near Smithville, Texas, originally described as *Nipadites burtinii* by Chester Arnold in 1952, deposited in the Paleobotany Collection of the Museum of Paleontology at the University of Michigan. The fossilization of this plant part in ironstone has preserved the angled ridges and the terminal knob so characteristic of *Nypa* fruits. Photo by Georg Oleschinski.

geological record. There have been new studies on the similarity between fossil and recent fruits and pollen of *Nypa*, as well as on their morphological distinctiveness from all other fruits and pollen.

In 1990, a fossil flora and fauna of middle Eocene age discovered on the Texas–Mexico border at Laredo yielded evidence of a brackish-water estuary with oyster reefs. Climatic and habitat analysis of the sharks, rays, fish, and 59 species of terrestrial vertebrates all pointed to this area being a tropical coastal environment. Yet, only the fossil pollen (see fig. 17.4) and fossil fruits (fig. 17.7) of *Nypa* confirmed it as a mangrove community. The Eocene habitat was then reconstructed as an estuary fringed by an extensive zone of *Nypa* palms. Behind the mangrove zone was a back-mangrove with ferns similar to the mangrove fern, known today as *Acrostichum*, and tropical trees characteristic of mangrove habitats today. Because mangrove palms have such specific climatic preferences today, it was possible to reconstruct likely Eocene temperatures and rainfall in what is dusty and dry Laredo today.

At the end of the Eocene, when the cooling of the planet began in earnest, *Nypa* disappeared from its estuarine habitat in North Amer-

Figure 17.7. A fossil fruit of *Nypa* from the middle Eocene of Laredo, Texas, that had just been freshly pried out of a 40-million-year-old rock. Photo by author.

Figure 17.8. The mangrove palm, *Nypa fruticans*, in Vietnam. Photo by Ming-Lee Tsay-Reichert.

ica. The intertidal niches in the New World were then taken over by *Rhizophora* and a whole host of other mangrove species that moved northward from Central America and the Caribbean into subtropical Florida. Thus, while the mangrove palm, *Nypa*, is restricted today to the coastlines of the Indian and Pacific Oceans (fig. 17.8), it once dominated the world's mangrove habitats. It is recognized as a survivor because it has continued to flourish for the 75 million years since the Upper Cretaceous.

FURTHER READING

Arnold, C. A. 1952. Tertiary plants from North America. 1. A *Nipa* fruit from the Eocene of Texas. *Palaeobotanist* 1:73–74.

Dowe, J., and R. Tucker. Last updated 2013. *Nypa fruticans.* PACSOA Palms and Cycads wiki. http://www.pacsoa.org.au/wiki/Nypa_fruticans.

Frederiksen, N. O. 1981. Middle Eocene to early Oligocene plant communities of the Gulf Coast. In *Communities of the Past*, ed. J. Gray and W. B. N. Boucot, 493–549. Stroudsbury, PA: Hutchinson Ross.

Gee, C. T. 2001. The mangrove palm *Nypa* in the geologic past of the New World. In Life on the Edge: Past and Future of Mangroves, ed. N. C. Duke and A. E. Schwarzbach. Special issue, *Wetlands Ecology and Management*, 9:181–94.

Westgate, J. W., and C. T. Gee. 1990. Paleoecology of a middle Eocene mangrove biota (vertebrates, plants, and invertebrates) from southwestern Texas. *Palaeogeography, Palaeoclimatology, Palaeoecology* 78:163–77.

INDEX

Page numbers in **bold** indicate figures.

Acrostichum (mangrove fern), 241
Adiantum (maidenhair fern), 117
Adiantum capillus-veneris (common maidenhair fern), 117
Agathis (kauri): cones, 96–97; distribution of, 140; images, **79**, **137–38**, **146**; morphology, 137, **138**; number of species, 140; taxonomy, 131, 137, 138; threats to, 145–46
Agathis microstachya (bull kauri), **79**
Agathis ovata (mountain kauri), 145, **146**
Agathis robusta (Queensland kauri), **138**
Aglaomorpha quercifolia (oakleaf basket fern), **79**
Allen, Jan, 139
American white water lily. See *Nymphaea odorata*
ancestral angiosperm flower, **191**, 196–97
Angiopteris (king fern), 40, **81**
angiosperms, 191–98; characteristics of, 194–96; dominance of, 8–9, **11**, 191, 197; food uses, 191, 198; fossils and evolutionary history, 8–9, **11**, 92, 192–94, 196–98, 209–14, **210**, **212**, 226–29, **228**, **235**, 235–42, **239–41**; pollen and pollination, 191–92, 195–96, 198; wood, 195. *See also* flowers; fruits
Annularia stellata, **68**
Antelope Island (Utah), 26, **27**
Araucaria (araucarias): cones, 96–97, **98**, **132–33**, 136, **137**, 143–44, **143–45**; distribution of, 9, 140; fossils and evolutionary history, 8, **11**, 140–46, **142–45**; height, 132–33; herbivory of, **71**, 136; images, **130**, **132–35**, **137**, **142–45**; leaves, **130**, 131, **132**, 143, **144**; morphology, 9, 131–32, **132**, 136; number of species, 140; seeds, 144; taxonomy, 131, 137, 138; threats to, 133, 145–46; uses, 132, 133–35, 146; wood, 143–45, **145**

Araucaria angustifolia (Paraná pine), 132–33, **134**
Araucaria araucana (monkey puzzle tree), 131–32, **132–33**, 136
Araucaria bidwillii (bunya pine), 133–35, **135**, 136, **137**
Araucariaceae, 131–46; distribution of, 140, 157; fossils and evolutionary history, 140–46, **142–45**; number of species, 140. See also *Agathis*; *Araucaria*; *Wollemia nobilis*
Araucaria cunninghamii, **98**
Araucaria goroensis, **130**, 140
Araucaria mirabilis, 143, **143**
Archaeopteris, **10**, 38
Archaeosigillaria, **52**
Archaeosperma, 91
Arecaceae, 233
Arnold, Chester, 240
Asian royal fern, **82**, 83
Asterophyllites longifolius, **68**
Avicennia, 236
Azolla (duckweed fern), **80**
Azolla rubra (red azolla), **80**
Baiera gracilis, **124**

bald cypress, 152, 170, 177
bark: eucalyptus, **105**; lycophytes, 40, 53; redwoods, 161, 170–71; *Sciadopitys*, 178; totara, 155; *Wollemia nobilis*, 137, 139
basket ferns, **79**
bennettitaleans, 157
beta-myrcene, 110
blue-green algae. *See* cyanobacteria
blue lotus, **202**
bogmoss, 45–46, 47
Bolivian water lily. See *Victoria boliviana*
bonsai, 106, **148**, 155
brandy bottle, 207
Buddhist pine. See *Podocarpus macrophyllus*
bull kauri, **79**
bunya pine. See *Araucaria bidwillii*
burls, 168

Caesalpinia pulcherrima (dwarf poinciana), **195**
Calamites, 40, 65–67, **67–68**
Calamophyton, 34–35, **35**
Calamophyton primaevum, **35**
Callixylon. See *Archaeopteris*
Carboniferous: and cycads, 101; and ferns and tree ferns, 34–35, 40, 81–83; and gymnosperms, 40–41, 54; and lycophytes, 38–40, 51–54, **55**; and sphenophytes, 40, 43, 65–68, **67–68**, 72; swamp forests, **10**, 38–40, **39**; in timeline, **10**; and tree height, 38–40, 42
carpels, 194, 195, 196–97, 225
Carus, Carl Gustav, 5, **5**
Cenozoic: and angiosperms, 9, 197; and *Araucaria*, 140, 145; and *Ginkgo*, 123, 126–27; and *Metasequoia*, 172–73; and *Nelumbo*, 227, 229; and *Nypa*, 238; as part of Phanerozoic, 8; and podocarps, 157–58; and *Sciadopitys*, 180–81, 182; and seed ferns, 93; in timeline, **11**; and water lilies, 211–12, 213
central plate, lotus, 223–24, 227, 228, **228**
Cerro Cuadrado Petrified Forest, **11**, 143, **143**
Chaney, Ralph W., 166–67
Cheirolepidiaceae, 157
Cheng, Wan-Chun, **13**, 164, 166, 170
chestnut dioon, 108
Chinese podocarp, **157**
Chinese swamp cypress, 163–64, 170
chlorophyll, 26–29, 117, 153
Chu, K. L., **13**
cinnamon fern, 83, 84–85
cladodes, 180–81
cladoxylopsids, **10**, **35**, 35–37, 40, 64
Clifton, Lake, **26**
clubmosses. *See* lycophytes (clubmosses)

coast redwood. See *Sequoia* or *S. sempervirens*
collecting ferns, 75–76, **76**
color: cyanobacteria, **18**, 26–29, **28**; cycad seeds, 94, 108, **109**; flowers, variety in, 194, **195**; *Ginkgo* leaves, 115, **116**, 117, 118, **119**; *Metasequoia* leaves, 167, **168**; *Nelumbo*, 221–22; water lilies, 201, 202, 205, 206, **207**, 207–8
common clubmoss. See *Lycopodium* or *L. clavatum*
common mossfern, 80
cones, 89–99; araucarias, 96–97, **98**, **132–33**, 136, **137**, 143–44, **143–45**; conifers, 93, 95–99; cycads, 93–95, **94**, 104, 108, **109**, 110; *Equisetum*, 64, **65–66**, 72, 93; fossils and evolutionary history, **11**, 89, 93–99, **143–45**, 155–56; *Ginkgo*, 118, **118**; kauris, 96–97; lycophytes, 47, **54**, 56, **56**; *Lycopodium*, 47; *Metasequoia*, 161–63, **162**, 164–65, 168, 170, 172; morphology, 93, 96–99, **99**; orientation of, 97; podocarps, 155–56, **157**; redwood, 168, **170**; *Sciadopitys*, **11**, 178, **178–79**, 180, 184, **186**; serotinous, 95, **97**; shatter, 136, 144, **144**; with spores, 93; *Wollemia nobilis*, 96–97, **98**, **141**; woody, 14
conifers: deciduous, 167–68, **168**, 170, 173; fossils and evolutionary history, **10**, 57; nutritional value of, 70, **71**; parasitic, 152–54, **153**, 158–59; seed evolution, 91, 92; tree morphology, 117, 136; wood, **10**, 42, 57. See also *Agathis*; *Araucaria*; podocarps; *Sciadopitys*; *Wollemia nobilis*
Cookson, Isabel, 34
Cooksonia, **10**, 33–34
coral tree, 152–54, **153**, 158–59
Cordaites, 40–41, 66, 82
corms, 56, **56**
Cotham (Landscape) Marble, 22–24
creeping strawberry pine, 151, **151**, 158
Cretaceous: and angiosperms, 8, 191, 192, 197–98, 209; and cycads, 102; and *Ginkgo*, 126; and *Metasequoia*, 171; and *Nelumbo*, 226–27, 229; and *Nypa*, 10–12, 237, 238, 242; and podocarps, 156, 157; and *Sciadopitys*, 180; and seed ferns, 93; in timeline, **11**; and water lilies, 209, 210–11, 213
Cretaceous-Paleogene (K/Pg) extinction event, **11**, 102
crossfield pitting, **182**
cryptograms, 89
Cupressaceae, 171
cupule, 91
cuticle, 14, 33, 126
cyanobacteria, 17–29; fossils and evolutionary history, 8, **10**, 17–20; images, **18**; layered colonies of, 20–22, **21**; photosynthesis by, 17–19, 21–22, 26–29; in stromatolites, 9, **12**, **16**, 20–29, **21**, **23**, **25–27**
Cyathea, 35, **79**, **104**
cycads, 101–13; cones, 93–95, **94**, 104, 108, **109**, 110; distribution of, 105; diversification of, 111–12; fertilization and pollination, 91–92, 93–94, 110; fossils and evolutionary history, 8, **10**, 57, 91, 101–2, 109, 111–12; habitat, 105, **105**, 112; height, 102; herbivory of, **71**, 106–8, 109–10, 113; images, **3**, **94**, **100**, **102–7**, **109**, **111–12**; as known living fossil, 3; leaves, 93, 102, 104, 105–8, 110–11, **111**, 112–13; longevity of, 106, 113; morphology, 102–5, 110–11, 112–13, 117–18; nutritional value of, 70–71, **71**, 113; seeds, 91–92, 93–94, 95, 104, 107–11, **109**, **111**; uses, 106, **106**, 107, 108
Cycas circinalis (queen sago), 102, **103**
Cycas revoluta (sago palm), 94–95, **100**, **102**, **106**, 107–8, 110–11, **111**
Cylindrospermum, **18**
cypress, 171. *See also* bald cypress; Chinese swamp cypress
Czekanowskia, **125**, 125–26

Dacrycarpus, 155
Dacrydium cupressinum (rimu), **150**
Darwin, Charles, 6, **6**, 192
date palm, **104**
dawn redwood. See *Metasequoia* or *M. glyptostroboides*
deciduous conifers, 167–68, 170, 173
decorative and horticultural use: araucarias, 132, 133–35, 146; cycads, 106, **106**; ferns, 75–77, **76**, 85; *Ginkgo*, 119–22; lycophytes, 47–50; *Metasequoia*, 167, **168**; *Nelumbo*, 217, **219**, 221, 223, 224–26, 229; podocarps, **154**, 154–55; *Sciadopitys*, 178; water lilies, 201–2
Devonian: and cladoxylopsids, 37, 64–65; and dichotomous venation, 117, 127; and Gilboa Fossil Forest, 36; and height, 34–38, **35**, 64–65; and lycophytes, 37–38, 51–53, **52**; megaphyll/microphyll split in, 51; seed development in, 91; in timeline, **10**
Dicksonia antarctica, **74**
Dicranopteris linearis (Old World forked fern), **78**, 79
dinosaurs, **11**, 70–71, 72, 101, 113
Dioon edule (chestnut dioon), 108
Dioon spinulosum (giant dioon), **104**
Diphasiastrum digitatum (fan clubmoss or trailing pine), 47–49, 50
distribution. *See* relictual distribution and refugia
diversity in living fossils, as low, 7, 13–14
djeeri, 102–4
double fertilization, 196
Drynaria quercifolia. See *Aglaomorpha quercifolia*
duckweed fern, 80
Dutch rush. See *Equisetum hyemale*
dwarf poinciana, **195**

Eastern Cape blue cycad, 107, **107**
elaters, 69, **69**
Elatocladus planus, **158**
Elksinsia polymorpha, 91
Encephalartos horridus (Eastern Cape blue cycad), 107, **107**
Encephalartos laurentianus (Kwango giant cycad), 102
endosperm, 91, 196, 237
Eocene: and *Metasequoia*, 172; and *Nelumbo*, 227; and *Nypa*, 12, **235**, 237–41, **239–41**; and *Sciadopitys*, 181, 182; in timeline, **11**; and waterlilies, 212–14, **214**
Eospermatopteris, 36
epiphytic habit: ferns, 79, **79**; lycophytes, 45, 47
Equisetites, **68**, 69
Equisetites columnaris, **68**
Equisetum, 61–73; cones, 64, **65–66**, 93; as cryptograms, 89; fossils and evolutionary history, 8, 9, **11**, 64–73, **66–69**; habitat, 61, 64, 89; height, 40, 42–43, 64–65, 70; images, **42**, **60**, **62–63**, **65–69**; leaves, 64, 67, **68**; and low diversity, 14; morphology, 9, 64, 67, **69**, 69–70; number of species, 64; nutritional value of, 70–71, **71**; silica in, 42–43, 62–63, **63**, 72; spores and reproduction, 64, **66**, 69, **69**, 92, 93; uses, 61–62, 72
Equisetum arvense (field horsetail), **42**, **62–63**, 63, 64, **65–66**, 72
Equisetum giganteum (giant horsetail), 43, 72
Equisetum hyemale (scouring or Dutch rush), **60**, 61–62
Equisetum sylvaticum (wood horsetail), **63**
Estinnophyton wahnbachense, **52**
European water clover, 80
evolution: first organisms, 17; overview of plant evolution, 8–15, **10–11**; split between free-sporing and seed plants, **51**, 92; split between microphylls and megaphylls, 51, **51–52**; of tree habit, 10, 14, 33–42, 65–66

Exnelumbites, 227
extinction: Cretaceous-Paleogene (K/Pg) event, **11**, 102; low rates in living fossils, 7; Permian-Triassic event, **10**, 55, 57; Pleistocene, 127; present-day pressure of, 75, 113, 145–46, 158–59, 173. *See also* rediscovery of species thought extinct

Falcatifolium taxoides, **153**, 153–54
false rose of Jericho, 46
fan clubmoss, 47–49
ferns and tree ferns, 75–85; collecting, 75–76, **76**; as cryptograms, 89; *vs.* cycads, **104**, 104–5; fiddleheads, 80, **81**; fossils and evolutionary history, 8, 9, **10–11**, 40, 57, 81–85, 92–93, 157; habitat, 76–77, **77**, 89; height, 34–35, 40, 79, **79**, 85; images, **39**, **41**, **74**, **76–82**, **84**, **92**, **104**; leaves and fronds, 40, **66**, 75, **77–78**, 77–79, 80–81, **82**, 83–84, **84**, 104; morphology, **66**, 77–79, 80, **81**, 83, **84**, 85; number of species, 81; nutritional value of, 70, **71**; root mantles, 34–35, 40, **41**; spores and reproduction, **51**, **66**, 92, 105; stem anatomy, 83; uses, 75–77, **76**, 80, 85
fertilization: double, 196; fossils and evolutionary history, 91
field horsetail. See *Equisetum arvense*
fire: and araucarias, 145; and flower fossils, 194; and height, 37; and lycophytes, 40, 50; and podocarps, 159; and serotinous cones, 95, **97**; and *Wollemia nobilis*, 140
firmoss, 45, 47
flowers: first, 194, 196–97, **210**; fossils and evolutionary history, **11**, 14, 15, 67, 191–98, 209–11; lotuses, **202**, **216**, 217, **218–19**, 220–23, 225–26; morphology, 194–95, **195**; *Nypa*, 235; water lilies, **11**, **195**, 196–97, **200**, 201–10, **203–4**, **210**, **217–18**, 223
fluorescence, 117
food, animal. *See* herbivory
food, human: angiosperms, 191, 198; cycads, 108; *Equisetum*, 72; ferns, 80; *Ginkgo*, 115, 120, **121**; *Nelumbo*, 217, 218, 220, **220–22**, 225–26, 229; *Nypa*, 236–37; water lilies, 207
forked ferns, **78**, 79
fragrant water lily. See *Nymphaea odorata*
Frederiksen, Norman, 239
fruits: dispersal of, 198; fossils, 194, 212–13, 226, 227, 237–41, **239–41**; *Ginkgo*, 118–19, **119**, 126; *Nelumbo*, 217, **218–19**, **222**, 222–23, 225, 227; *Nypa*, **234**, 234–35, 236–37, 238–41, **239–41**; palms, 104; water lilies, 207, **208**, 209, 213
Fu, Shu-Hsia, 164

Gee, Carole T., **187**
geological time, 8–12, **10–11**
giant dioon. See *Dioon spinulosum*
giant horsetail. See *Equisetum giganteum*
giant redwood/sequoia. See *Sequoiadendron* or *S. giganteum*
Gilboa Fossil Forest, 36
ginkgo or *Ginkgo biloba*, 115–29; autumn gold cultivar, 119–20; at Bonn University Botanic Gardens, **119**, 120–21; cones, 118, **118**; fall leaf coloration, 115, **116**, 117, **119**; fossils and evolutionary history, **4**, 8, **10–11**, 123–29, **124–25**; habitat, 122, 126, 127; height, 117; images, **3**, **4**, **114**, **116**, **118–19**, **123–25**; leaves, 115–17, **116–17**, 119, **119**, **124–25**, 124–26, 127–28; lignotubers, 122, **123**; as living fossil, 3, **4**, 115, 127–29; longevity of, 120, 121–22; and low diversity, 13–14, 129; morphology, 115–18, 127, 129; nutritional value of, 70–71, **71**; Old Lion at Royal Botanic Kew, 120; popularity of, 115; reproduction, 118–21, 122; seeds and pollination, 92, 118, **118–19**, 120, **121**, 123, 126; uses, 115, 119–21

Ginkgo adiantoides, **4**
ginkgo berries, 118–20, **119**, 126, 127
ginkgofluor, 117
Ginkgoites, 125, **125**
Ginkgoites cascandensis, **125**
ginkgo nuts, 118, 120, **121**
Ginkgo Petrified Forest State Park, **11**, 127, **128**
ginkgophytes, **10**, 123–29, **124–25**
Glyptostrobus pensilis (Chinese swamp cypress), 163–64, 170
gnetophytes, 118
golden larch, 170
Goldring, Winifred, 36
Great Dying, **10**, 55, 57
Great Oxygenation Event, **10**, 19–20
guard cells, 33, 62
gymnosperms: Mesozoic, 192; native to Sri Lanka, 103; and shifts in flora, 8–9, **10–11**, 54–55, 57, 191

Hayashi, Shoji, **187**
height. *See* tree habit
herbivory: of araucarias, **71**, 136; of cyanobacteria, 22; of cycads, **71**, 106–8, 109–10, 113; in defenses overview, 14; of *Equisetum*, 70–72, **71**; of *Ginkgo*, **71**, 120; and height, 37; nutritional comparison, **71**; of podocarps, 159
Hiroshima, 122
Hohenheim Feeds Evaluation Test, 70–71, **71**
hoop pine. See *Araucaria cunninghamii*
Hope's cycad, 102
horsetail. See *Equisetum*
Hsueh, Chi-Ju, 164
Hu, Hen Hsu, 164–66, 170
Huperzia (firmoss), 45, 47
Huperzia phlegmarioides (tassel fern), 47, **48**
Hwa, Ching-Tsan, **13**, 166
Hypericum (St. John's wort), **195**

interrupted fern, 83–84, 85
Ipomoea tricolor (morning glory), **195**
Isoetes (quillworts) or *I. melanospora*, 45, 46, 47, **49**, **51**

jack pine, 95
Japanese umbrella pine. See *Sciadopitys* or Sciadopityaceae
Jones, Wyn, 139
Jurassic: and *Araucaria*, 9, 136, 142–43, **143–45**; and cycads, **112**, 113; and *Czekanowskia*, **125**, 125–26; and *Equisetum*, **68–69**, 70, 71; and ferns, 84, **84**; and *Ginkgo*, 9, 71, 123, **124**, 125, 126; and lycophytes, **58**; and podocarps, 157, **158**; and *Sciadopitys*, 180; in timeline, **11**
Jurassic Salad Bar (Utah), **125**, 125–26

kauri. See *Agathis* (kauri)
king fern, 40, **81**
kōyamaki. See *Sciadopitys* or Sciadopityaceae
K/Pg extinction event, **11**, 102
Kwango giant cycad, 102

Laguncularia, 236
Landscape Marble, 22–24, **23**
Lang, William Henry, 34
Larix (larch), 170
Lazarus taxa, 12. *See also* rediscovery of species thought extinct
least water lily, **208**
leaves: angiosperms, 70, 195; evolutionary split of microphylls and megaphylls, 51, **51–52**. *See also* venation
Lepidodendron (scale tree), 38–40, 53, **54–55**
Lepidozamia hopei (Hope's cycad), 102
Lesvos orchid, **195**
lignotubers, 122, **123**
Lindlar forest (Germany), 34, 36
Linnaeus, Carl, 63, 206

living fossils: characteristics, concepts, and definitions, 3–15; rediscovery of species thought extinct, 4, 7, 12–13; as term, 5, 7–8. See also *Agathis*; cyanobacteria; cycads; *Equisetum*; ferns and tree ferns; ginkgo or *Ginkgo biloba*; lycophytes; *Metasequoia* or *M. glyptostroboides*; *Nelumbo* or *N. nucifera*; *Nypa* or *N. fruticans*; podocarps; *Sciadopitys* or Sciadopityaceae; water lilies; *Wollemia nobilis*
locules, lotus seed, **222**, 222–23
lodgepole pine, 95, **97**
lotus, blue or star, **202**
lotus, sacred. See *Nelumbo* or *N. nucifera*
lotus effect, 224, **225**
lycophytes (clubmosses), 45–58; bark, 40, 53; collecting, 76; cones, 47, **54**, 56, **56**; as cryptograms, 89; diversity of, 50–51; epiphytic habit, 45, 47; fossils and evolutionary history, 8, **10**, **39**, 50, **51–52**, 51–58, **54–56**, **58**; habitat, 45–46, 57, 58, 89; height, 37–40, 53, 54; images, **39**, **44**, **46**, **48–49**, **52**, **54**, **57–58**; leaves, 45, **49**, 51, **51–52**; morphology, 45, 47, 58; spores, 47, 49–50, 56, 92; as spore spike, 50; uses, 47–50
Lycopodiella, 47. *See also* bogmoss
Lycopodites falеatus, **58**
Lycopodium, **44**, **46**, 47, 50
Lycopodium or *L. clavatum* (common clubmoss), **44**, **46**, 50
lycopodium powder, 50

Macrozamia dyeri (djeeri), 102–4
Macrozamia johnsonii, 108, **109**
Macrozamia lucida, 110
Macrozamia miquelli, 110
macrozamin, 108
maidenhair fern, 117
maidenhair tree. *See* ginkgo or *Ginkgo biloba*
mangrove, 41, 231–32, **232**, 236, 237, 242
mangrove fern, 241
mangrove palm. See *Nypa* or *N. fruticans*
Marattia or Marattiaceae, 40
Marsileaceae, 79–80, **80**
Marsilea quadrifolia (European water clover), **80**
Meganeura, 53
megaphylls, 51, **51**
megasporophylls, 94–95, 110–11, **111**
Merrill, Elmer D., 166
Mesozoic: and angiosperms, 8, 192; and *Araucaria*, 8, 9, 140–41, 143, 145, 146; and cycads, 4, 101–2, 111, 112–13; and *Czekanowskia*, 126; and *Equisetum* and sphenophytes, **68**, 69–73, **71**; and ferns, 83, 93; and *Ginkgo*, 123–24, 129; and *Nelumbo*, 8; and *Nypa*, 8, 237; and podocarps, 8, 155–59; recovery from extinction events in, 57; and *Sciadopitys*, 180, 182, 186; in timeline, **11**; and water lilies, 8
Metasequoia or *M. glyptostroboides* (dawn redwood), 161–73; cones, 161, **162**, 163, 164, 168, **170**; deciduous habit, 163, 167–68, 170, 173; distribution of, 127, 167, **168**, **170**, 171, 173; fossils and evolutionary history, **11**, 161–63, **162**, 168, 171–73; habitat, 167; height, 161, 168; images, **13**, **160**, **162**, **165**, **168–70**; leaves, 167–68, **168**, 170, 171; and low diversity, 14; morphology, 167–68; rediscovery of, 4, 12–13, 161–67, 173; uses, 167, **168**; wood and bark, 161, 170–71
Metasequoia occidentalis, **170**
Metasequoia Valley (Moudao, China), 163–64, 166–67
microbialites, 25–26, **26–27**
Microcachrys tetragona (creeping strawberry pine), 151, **151**, 158
microCT, 98–99, **99**

microphylls, 51, **51–52**
Middle Eocene Climatic Optimum, **11**, 238
Miki, Shigeru, 161–63, 164
Miocene: and cycads, 111; and *Ginkgo*, 127; and *Metasequoia*, 161, **162**, 173; and *Sciadopitys*, 9, 181–83; in timeline, **11**
Miroviaceae, 180
mitochondria, 20
Monet, Claude, 201–2
monkey puzzle tree, 131–32, **132–33**, 136
monocots, 237
Mono Lake (California), 26, **28**
morning glory, **195**
morphological characteristics of living fossils, 4–5, 6, 9
mossfern, common, **80**
Moudao, China, 163–64, 166–67
mountain kauri, 145, **146**

Nelumbo or *N. nucifera* (sacred lotus), 217–29; distribution of, 229; diversity of, 229; flowers, **216**, 217, **218–19**, 221–22, 225; fossils and evolutionary history, 8, **11**, 226–29, **228**; fruits, 217, **218–19**, **222**, 225, 226, 227; images, **216**, **218–22**, **225**, **228**; leaves, 217, 218, **218–19**, 223–24, **225**, 226–28, **228**; morphology, 198; roots and rhizomes, 217–20, **220–21**, 223–27; seed pods, **218**, **222**, 222–23, 227; seeds, 91, 217, 218, **222**, 222–23, 225, 226, 227; uses, 217, 218, **219–22**, 220, 223, 224–26, 229; *vs.* water lilies, 205, 220
Nelumbo lutea (yellow lotus), 221
Nelumbonaceae, 220–21, 227–29
New Caledonia, 140, 145–46, 149, 151–52, **152–53**, 159
New Caledonia corkwood, 151–52, **152**
Nipadites burtinii, **240**
Noble, David, 138–39
nodal diaphragms, **69**, 70
Notonuphar, 213
Nuphaea, 213, **214**
Nuphar (yellow water lily): distribution of, 212, 213; flowers, 207–9, **208**, 223; fossils and evolutionary history, **11**, 209–14; fruits, **208**, 209, 212–13; leaves, 208, **208**, 223; morphology, 198, 207–9; pollination, 207–9; seeds, 211
Nuphar lutea (yellow pond lily, spatterdock, or brandy bottle), 207
Nuphar pumila (least or small water lily), **208**
Nymphaea (white water lily): flowers, **195**, 196–97, **200**, 201–3, **202–4**, 209–11, **210**, 223; fossils and evolutionary history, 209–14, **210**, **212**; habitat, 203–4; images, **200**, **202–4**, **207**, **210**, **212**, **214**; leaves, **200**, 202–5, **203–4**, **207**, **212**, 223; morphology, 198, 202–4; pollination, 206–7
Nymphaea alba (Old World white water lily), 206
Nymphaea calophylla, 211, **212**
Nymphaea lignitica, 228, **228**
Nymphaea nouchali (blue or star lotus), **202**
Nymphaea odorata (American white water lily), **200**, **204**, 205–7
Nypa or *N. fruticans* (mangrove palm), 231–42; flowers, 235–36; fossils and evolutionary history, **11**, 12, **235**, 236, 237–42, **239–41**; fruits, **230**, **234**, 234–41, **239–41**; habitat, 197, 232–34, **233**, 236; images, **230**, **234–35**, **239–41**, **242**; leaves, 234, 236; and low diversity, 14; morphology, 198, 233–34; pollen, **235**, 235–36, 237–40, 241; seeds, 237; uses, 236–37

oakleaf basket fern, **79**
Odaigahara, Mount (Japan), 183–86, **184–86**, **187**

Old World forked fern, **78**, 79
Old World royal fern, **78**, 83
oncolites, 21
Ophrys lesbis (Lesvos orchid), **195**
Osmundaceae, 9, **78**, **82**, 83, 84, **84**, 85
Osmunda claytoniana (interrupted fern), 83–84, 85
Osmunda japonica (Asian royal fern), **82**, 83
Osmundales, 83–85
Osmunda regalis (Old World royal fern), **78**, 83
Osmundastrum cinnamomeum (cinnamon fern), 83, 84–85
ovules: cycads, 93–94, 110–11, **111**; *Ginkgo*, 118, 122; podocarps, 155; seed cones, 95. *See also* seeds

Paleocene: and *Metasequoia*, **170**, 172; in timeline, **11**
Paleozoic: and cycads, 8; and ferns and tree ferns, 8, 40, 93; and *Ginkgo*, 8; and lycophytes, 8, 53, 58; and podocarps, 159; and sphenopytes and *Equisetum*, 8, 68, **68**; in timeline, **10**
palms: culinary uses, 108; in Mesozoic, 8; in Miocene, 181; similar appearance to cycads, 4, 101, 102, **104**, 104–5. See also *Nypa* or *N. fruticans* (mangrove palm)
Paraná pine. See *Araucaria angustifolia*
Parasitaxus usta (coral tree), 152–54, **153**, 158–59
Permian: and cycads, 101; ferns and tree ferns, **41**, 83; and *Ginkgo*, 123; and lycophytes, 54–57; Permian-Triassic extinction event, **10**, 55, 57; and podocarps, 157; rise of gymnosperms, 54; and sphenophytes, 68–69; in timeline, **10**
Permian-Triassic extinction event, **10**, 55, 57
petals: first hypothetical flower, 196–97; *Nelumbo*, 217, **218–19**, 221, 222, 226; *Nuphar*, 207–8, **208**, 209, 212–13; variety in, 194, **195**; water lilies, **191**, 197, **200**, 201, **202–4**, 206, **207**, 208, 209, **210**, 213, **214**
petiolar diaphragms, 224
Petrified Forest (Arizona), **11**, **142**, 142–43
Phoenix dactylifera (date palm), **104**
photosynthesis: with cladodes, 180–81; by cyanobacteria, 17–18, 21–22, 26–29; in *Equisetum arvense*, 64, **65**; fossils and evolutionary history, 14; and height, 37; and parasitic conifers, 153; pigments and color in, 26–29; and volcanic eruptions, 57; and water lilies, 205
phycobilin, 29
pigments, 26–29, **28**, **116**, 117
pines, cones, 93, 95–99, **96–97**, **99**
Pinus contorta (lodgepole pine), 95, **97**
Pinus pinea (umbrella pine), **99**, 177, **178**
Pleistocene: and cycads, 109, 112; and *Ginkgo*, 127; and *Metasequoia*, 161, 171, 173; in timeline, **11**
Pleistocene megaherbivore fauna, 109
Pleuromeia or *P. sternbergii*, **10**, 55–57, **56**
Pliocene: and cycads, 112; and *Metasequoia*, **162**; and *Sciadopitys*, 180; in timeline, **11**
podocarps, 149–59; distribution of, 149–50, 157–58; diversity of, 149; fossils and evolutionary history, 8, **11**, 140, 143, 155–58, **158**; habitat, 149–52, **150–52**; height, 150–52, 159; herbivory of, 159; images, **148**, **150–54**, **156–58**; leaves, 150, 153, **153**, 155–56, **158**; morphology, 149, **150–54**, 150–56; nutritional value of, **71**; parasitic, 152–54, **153**; seeds and seed cones, **153**, 155–56, **157**; threats to, 145–46, 158–59; uses, **154**, 154–55
Podocarpus chinensis (Chinese podocarp), **157**

Podocarpus macrophyllus (Buddhist pine), **148**, **150**, **154**, 155
Podocarpus totara (totara), 155, **156**
pollen and pollen cones, 179, 181, 183; angiosperms, **11**, 192, 194, **195**, 196, 209, 211; angiosperms *vs.* gymnosperms, 192, 195–96; araucarias, **132–33**, 136, **137**; and beta-myrcene, 110; conifers, 91, 93, 95; and Cookson, 34; cycads, 91, 94, 110; fertilization, 91, 196; in fossil studies, 50; *Ginkgo*, 118, 120, 122; *Nymphaea*, 206, 208–9; *Nypa*, **11**, **235**, 235–36, 237–40, 241; pines, 95, **96**; podocarps, 156, 158; and pollen chamber, 91; *Sciadopitys*, 179, 181, 183; in stamens, 194; and thrips, 110; *Wollemia nobilis*, **141**
pollination: angiosperms, 191–92, 196, 198; cycads, 93, 110; flower color and size, 194; push-pull, 110; water lilies, 206–9
Precambrian: and cyanobacteria, 8, 20–21; in timeline, **10**
progymnosperms, 38
Proteales, 220–21
Protolepidodendropsis, 38
Psaronius or *P. infarctus*, 34–35, 40, **41**
Pseudoctenis or *P. ensiformis*, 112, **112**
Pseudolarix (golden larch), 170

Quaternary, **11**
queen sago. See *Cycas circinalis*
Queensland kauri, **138**
quillworts, 45, 46, 47, **49**, **51**

red azolla, **80**
rediscovery of species thought extinct: among living fossils, 4, 7, 12–13; *Ginkgo*, 122; *Metasequoia*, 4, 12–13, **13**, 161–67, 173; *Sciadopitys*, 180; *Wollemia nobilis*, 13, 138–39
red mangrove. See *Rhizophora stylosa* or *R. stylosa*
redwoods. See *Metasequoia* or *M. glyptostroboides* (dawn redwood); *Sequoiadendron* or *S. giganteum* (giant redwood); *Sequoia* or *S. sempervirens* (coast redwood)
relictual distribution and refugia: *Ginkgo*, 9, 127, 128, 129; in living fossils, 7, 9, 12; *Metasequoia*, 127, 167; *Sciadopitys*, 182, 183–86
religious uses, 121, 155, 175–77, 217
reproduction: asexual, 122, 226; cryptograms, 89; and seed evolution, 89–92. *See also* pollination; spores
resin, 95, 137, 145, **145**, 181
resurrection plant, 46
Retrophyllum minus (New Caledonia corkwood), 151–52, **152**
rhizoids, 33, 34
rhizomes: *Equisetum*, 64, 72; fossil Osmundaceae, 83, 84; *Nelumbo*, 217–20, **220–21**, 223–27; water lilies, 207, 211, 213
Rhizophora stylosa or *R. stylosa* (red mangrove), 41, **232**, 236, 237, 242
rimu. See *Dacrydium cupressinum*
roots: angiosperms, 195; buttress, 168, **169**; ferns, 77, 79, 80, 84; fossils and evolutionary history, 14, 33–34, 37, 56, 227; lycophytes, 56; *Metasequoia*, 168, **169**; *Nelumbo*, 217–20, **220–21**, 223–27; prop and stilt roots, **41**, **232**; root mantles and tree habit, 34–35, 37, 40, **41**; *Wollemia nobilis*, 139, 140
royal ferns. *See* Osmundaceae

sacred lotus. See *Nelumbo* or *N. nucifera*
sago palm. See *Cycas revoluta*
saguaro, **193**
salinity: and mangroves, 231–32; and *Nypa*, 232–33, 234, 236; of Shark Bay, 24–26; and stromatolites, 9, 22–26
Salvinia or Salviniaceae (watermoss), 79–80, **80**

Salvinia minima (common mossfern), **80**
Sander, Chris, **187**
Sander, Laura, **187**
Sander, Martin, **187**
Sander, Philipp, **187**
sarcotesta, 118–19, **119**. *See also* ginkgo berries
scale tree. See *Lepidodendron*
Schopf, Tom, 6–7
Schweitzer, Hans-Joachim, 165
Sciadopitys or Sciadopityaceae (Japanese umbrella tree), 175–87; bark, 178; cones, **174**, 178, **178–79**, 180, 184; distribution of, 9, 178–79, 181–82; diversity of, 14, 182; fossils and evolutionary history, **11**, 180–82, **182**; habitat, 178–79, 182; height, 175, 178; images, **174**, **176**, **178–79**, **182**, **185–87**; leaves, 177, 178, 180, **182**; morphology, 177, 178, 180; seeds, 180; uses, 175–77, **176**, 178, 179; visit to native habitat of, 183–87; wood, 177, 179, 180, 181, **182**
Sciadopitys verticillata, **174**, 177–82, **178–79**, **186**
scouring rush, 61–64. See also *Equisetum*
seed coats, 91, 94, 118, 226
seed cones. *See* cones
seed dispersal: *Araucaria*, 136; cycads, 108–10; *Ginkgo*, 120; *Nypa*, 235; podocarps, 155–56; shatter cones, 136, 144, **144**
seed ferns, 40, 57, 82, **92**, 92–93, 157
seeds: angiosperms, **90**; cycads, 91–92, 93–94, 95, 104, 107–11, **109**, **111**; fossils and evolutionary history, 8, **10–11**, 14–15, 38, 89–92, **162**; *Ginkgo*, 92, 118, 120, **121**, 123, 126; *Macrozamia*, **109**; *Metasequoia*, **162**; *Nelumbo*, 91, 217, 218, **218**, **222**, 222–23, 225, 226, 227; *Nypa*, 237; *Pinus*, **99**; *Podocarpus*, **157**; seed ferns, **92**; water lilies, **11**, 206–9, 211, 213
Selaginella (spikemoss), **46**, 47, **49**
Selaginella bryopteris, **49**
Selaginella lepidophylla (resurrection plant or false rose of Jericho), **46**
Sensō-ji Temple (Asakusa, Japan), 122
sepals, 194, 196, 197, 209, 213
Sequoia or *S. sempervirens* (coast redwood), 161, 163, 164, 168, 170–71, **172**
"Sequoia chinensis." See *Metasequoia*
Sequoiadendron or *S. giganteum* (giant redwood), 164, 168, 170–71, **172**
serotinous cones, 95, **97**
Shark Bay, Australia, **12**, **16**, 24–25, **25**, 27, **28**
shatter cones, 136, 144, **144**
Sigillaria, 40, 53, **54**
sika deer, 183, **184**
silica, 43, 62–63, **63**, 72, 127, **128**
Silurian: and lycophytes, 45, 51; in timeline, **10**; and vascular plants, 33–34
Silverman, Milton, 167
small yellow water lily, **208**
Sonneratia, 236
spatterdock, 207
sphenophytes, **10**, **39**, 40, 42–43, 64–73, **68**
spikemosses, 45, 46, 47, **49**, **51**, 76
spores: *Archaeopteris*, 38; Cladoxylopsida, 36; and cones, 93; *Equisetum*, 64, **66**, 69, **69**, 92–93; ferns, **66**, 92, 105; heterospory, 47, 56; lycophytes, 47, 49–50, 56, 92; *Pleuromeia*, 56–57; seed ferns, 92
sporophytes, 89
stag's horn clubmoss. See *Lycopodium* or *L. clavatum*
stamens, 194, 196, 201, 206, 209, 213, **218**
star lotus, **202**
stigmatic disks, 209, 213
St. John's wort, **195**
stomata, 14, 33, 62, 223
strobili. *See* cones
stromatolites, 9, **12**, **16**, 20–29, **21**, **23–27**

tannins, 171
tassel fern, 47, **48**

Taxodium distichum (bald cypress), 152, 170, 177
tepals, 196
terrestrialization, **10**, 33–34
thrips, 110
Todea barbara, **84**
topiary, **154**, 155
totara, 155, **156**
toxins, 107–8, 113
trailing pine, 47–49, 50
tree ferns. *See* ferns and tree ferns
tree habit, **10**, 33–43, 65–66
Tsukagoshi, Minoru, 183

umbrella pine (*Pinus pinea*), **99**, 177, **178**
umbrella pine, Japanese. See *Sciadopitys* or Sciadopityaceae

venation: angiosperms, 195; *Czekanowskia*, 126; dichotomous, 116–17; *Ginkgo*, 116–17, 127; lotuses, 226, 227, 228, **228**; megaphylls *vs.* microphylls, 51; water lilies, 211, 213
Victoria boliviana (Bolivian water lily), 205

Wang, Chan, 163–64
Ward, Nathaniel Bagshaw, 76
Wardian cases, 76
water ferns, 79–80, **80**
water lilies, 201–14; distribution of, 204, 205–6, 211–12, 213; flowers, **11**, **195**, 196–97, **200**, 201–10, **203–4**, **210**, **217–18**, 223; fossils and evolutionary history, 8, **11**, **191**, 209–14, **210**, **212**, 228; fruit, 207, **208**, 209, 213; habitat, 203, 211–12, 228; images, **200**, **202–4**, **207–8**, **210**, **212**, **214**; leaves, **200**, **203–4**, 203–5, **207–8**, 208, 210–11, **212**, 213, 223; morphology, 198, 204–5, 206–9, **208**; *vs.* *Nelumbo*, 205, 220; pollination, 206–9; seeds, **11**, 206–9, 211, 213; size, 205; stems, 203–4, **204**, 205, 223; uses, 201–2, 207. See also *Nuphar*; *Nymphaea*
watermoss, 80, **80**
white water lily. See *Nymphaea alba* (Old World white water lily); *Nymphaea odorata* (American white water lily)
Wollemia nobilis (Wollemi pine): bark, 137, 139; cones, 96, **98**, 141; genetic diversity of, 14, 139–40; height, 139; images, **98**, **141**; morphology, 137; as part of Araucariaceae, 131, 137; rediscovery of, 13–14, 138–40; seeds, 139; threats to, 140
wood: angiosperms, 42, 195; araucarias, 143–45, **145**; *Archaeopteris*, **10**, 38; *Calamites*, 65–67, **67**; conifers, **10**, 42, 57; *Cordaites*, 41; *Equisetum*, 43, 61–62; fossils and evolutionary history, 8, 60, 123, 127; *Ginkgo*, 42, 54; Ginkgo Petrified Forest State Park, 127, **128**; ginkgophytes, 123; lycophytes, 40; podocarps, 155; redwoods, 161, 170–71; sanding of, 61, 62; *Sciadopitys*, 177, 179, 180, 181; "unaltered", 181
wood horsetail. See *Equisetum sylvaticum*
Wu, Chung-Lun, 164

Yang, Lung-Hsing, 163
yellow lotus (*Nelumbo lutea*), 221
yellow pond lily, 207
yellow water lily. See *Nuphar*

Zamia furfuracea, **106**
Zamia pumila, 107–8